Urban and Maritime Transport XXVIII

WIT*PRESS*

WIT Press publishes leading books in Science and Technology.
Visit our website for the current list of titles.
www.witpress.com

WIT*eLibrary*

Home of the Transactions of the Wessex Institute.
Papers contained in this volume are archived in the WIT eLibrary in volume 212 of WIT Transactions on the Built Environment (ISSN 1743-3509).
The WIT eLibrary provides the international scientific community with immediate and permanent access to individual papers presented at WIT conferences.
Visit the WIT eLibrary at www.witpress.com.

TWENTY EIGHTH INTERNATIONAL CONFERENCE ON URBAN AND MARITIME TRANSPORT AND THE ENVIRONMENT

Urban and Maritime Transport 2022

Conference Chairmen

Giorgio Passerini
Polytechnic University of La Marche, Italy
Member of WIT Board of Directors

Stefano Ricci
Sapienza University of Rome, Italy

International Scientific Advisory Committee

Organised by

Wessex Institute, UK

Sapienza University of Rome, Italy

Sponsored by

WIT Transactions on the Built Environment

International Journal of Transport Development and Integration

WIT Transactions

Wessex Institute
Ashurst Lodge, Ashurst
Southampton SO40 7AA, UK

We would like to express thanks to all the conference Chairs and members of the International Scientific Advisory Committees for their efforts during the 2022 conference season.

Conference Chairs

International Scientific Advisory Committee Members 2022

Urban and Maritime Transport XXVIII

Editors

Giorgio Passerini
Polytechnic University of La Marche, Italy
Member of WIT Board of Directors

Stefano Ricci
Sapienza University of Rome, Italy

WIT*PRESS* Southampton, Boston

Editors:

Giorgio Passerini
Polytechnic University of La Marche, Italy
Member of WIT Board of Directors

Stefano Ricci
Sapienza University of Rome, Italy

Published by

WIT Press
Ashurst Lodge, Ashurst, Southampton, SO40 7AA, UK
Tel: 44 (0) 238 029 3223; Fax: 44 (0) 238 029 2853
E-Mail: witpress@witpress.com
http://www.witpress.com

For USA, Canada and Mexico

Computational Mechanics International Inc
25 Bridge Street, Billerica, MA 01821, USA
Tel: 978 667 5841; Fax: 978 667 7582
E-Mail: infousa@witpress.com
http://www.witpress.com

British Library Cataloguing-in-Publication Data

A Catalogue record for this book is available
from the British Library

ISBN: 978-1-78466-477-0
eISBN: 978-1-78466-478-7
ISSN: 1746-4498 (print)
ISSN: 1743-3509 (on-line)

The texts of the papers in this volume were set individually by the authors or under their supervision. Only minor corrections to the text may have been carried out by the publisher.

Preface

These Proceedings contain the revised versions of selected papers contributed to Urban and Maritime Transport 2022 (UMT22), the 28th International Conference on Urban and Maritime Transport and the Environment, delivered online due the Corona virus pandemic, in September 2022.

The Urban Transport conference was part of one of the most successful annual conference series the Wessex Institute has been organising. Launched 28 years ago by Professor Carlos Brebbia, the Institute's founder, it becomes soon the premier event of its type, convened in many popular European venues and attracting an international spread of delegates.

Meanwhile, the substantial growth of maritime shipping has resulted in large transported quantities around the world, creating a demand for innovative solutions for ports and fleets, well reflected by the success of the First International Conference on Maritime Transport, held in September 2019.

These two seemingly parallel themes are interconnected in the transportation and environmental management of coastal cities, dramatically affected by the positive and negative effects of land and sea traffic.

The enduring demand for better urban transport systems and the need for a healthier environment create a fertile environment for original ideas, innovative approaches and applications of advanced technologies, tests, and practical evaluations. In addition, there is a growing need for integration with ICT systems and applications to improve security and efficiency.

On its side, maritime transportation is strictly complementary with rail, road and air services as well as inland waterways in order to maximise its efficiency and respond rapidly to variable economic and political contingencies.

The variety of topics covered by the conference reflects the complex interaction of the transport systems with their environment and the need to establish integrated strategies. Such strategies are primarily aimed at achieving optimal socio-economic solutions while reducing the negative environmental impacts of transportation systems typically by interdisciplinary approaches. Therefore, the Conference will focus on multidisciplinary research and development, as well as case studies.

Urban mobility is a primary concern in sustainable urban development policies. Different forms of urban mobility have very different impacts on the urban environment and on the quality of the public spaces around them. Furthermore, the fulfilment of the right to mobility is sometimes jeopardized by competitiveness in the inter-urban intermodal market. In this context, typical topics that populate the proceedings are traffic planning, control and management, urban policy and governance, mobility and public spaces, logistics and urban freight distribution, positioning, route planning and monitoring, fleet management, vehicle technologies, socio-economic implications, guided systems, innovative electric systems, safety and risk analysis, human behaviours, modelling and simulation, infrastructure development, ITS, education and training.

Meanwhile, one of the key issues of maritime transport is the amplification of effects generated by the big quantities of goods and the dimensions of investments in infrastructures and fleets needed to meet the demand and plan effective freight and passengers transport services. In this context, the challenge of technological and operational innovation is to setup the most useful solutions to effectively deal with such crucial problem. Nevertheless, economic and political instabilities, in the globalised world, are generating a geographically jeopardized alternation of development and recession, which often requires extreme performances in terms of capacity, speed, energy consumption, environmental and social sustainability, as well as economic efficiency. Therefore, interdisciplinary researches characterise the key topics and fosters the necessity of meeting opportunities for knowledge exchange among concerned researchers and operators.

The Editors wish to acknowledge the authors, the members of the Scientific Committee, the Referees, the Institutional Partners, who supported the Conference, and Marta Graczyk, of the Conference Secretariat, who did an outstanding job of coordination together with the WIT Press staff, Isabelle Rham, in particular.

Finally, the Editors, the Chairmen, the Members of Scientific Committee and of the Wessex Institute wish to remember the late Professor Carlos Brebbia, founder of Wessex Institute and of WIT Press, who foresaw and established this series of meetings.

The Editors, 2022

Contents

Section 1: Traffic planning, control and management

Shuttles & Co: Influencing autonomous shuttle buses using V2X communication
Christian Wille & Jan Trumpold 3

Big data in smart transportation planning: Towards an applicable smart socio-economic transportation platform in historical mega cities
Reem Bakri, Zeyad El-Sayad & Ali F. Bakr 13

Algorithm for planning faster routes in urban networks with time-dependent arcs and the possibility of introducing waiting periods at nodes
Francisco A. Ortega, Guido Marseglia, Juan A. Mesa & Ramón Piedra-de-la-Cuadra 25

Sustainable mobility, transit-oriented development, and travel behavior: A travel pattern analysis
Aya Hasan Alkhereibi 37

Section 2: Mobility in the urban environment

Concept of accessibility in sustainable transport: Criteria and perspectives
Hoda Pourramazani & Jose L. Miralles-Garcia 49

Landuse planning for electric vehicle infrastructure
Gargi Ghosh 61

Urban transport infrastructure for people with disabilities in the city of Tshwane, South Africa: An observation audit
Babra Duri & Rose Luke 73

Evaluating the accessibility of a sustainable city: Case study of Masdar City, UAE
Praveen Kumar Maghelal, Xiangyu Li & Anoud Qahtani 87

Section 3: Ports management and operation

Best practices exchange in sea–rail intermodality:
A case study of the ports of Izmir, Turkey, and Trieste, Italy
Zeynep Pehlevan & Stefano Ricci 99

Methodology framework for prioritisation of renewable energy sources in port areas
Sofia Agostinelli, Mehdi Neshat, Meysam Majidi Nezhad, Giuseppe Piras & Davide Astiaso Garcia 113

Emerging ICT and port community systems:
A survey of scientific literature
Giuseppe Musolino, Girolamo Peda & Francesco Russo 123

Section 4: Maritime and railways transportation

Significance and current regulations of external airborne noise from ships
Luka Vukić, Ivan Peronja & Nikola Mandić 139

Developing accountable maritime transport and port organizational structures in Arab countries
Khaled El Sakty & Alia Emad Islam 149

Main causes of cargo fire incidents in ro-ro spaces:
An analysis over the last 25 years
África Marrero, Paco Gasparín, Francisco Rodero & Pablo Sanz 161

MOSES Project: Enhancing short sea shipping with automated technologies
Mercedes de Juan, Ignacio Benitez & Jorge Marcos 173

M/V Ever Given: Legal assessment of the causes and consequences of the accident
Michael Boviatsis 185

Author index 197

SECTION 1
TRAFFIC PLANNING, CONTROL AND MANAGEMENT

SHUTTLES & CO: INFLUENCING AUTONOMOUS SHUTTLE BUSES USING V2X COMMUNICATION

CHRISTIAN WILLE & JAN TRUMPOLD
German Aerospace Center (DLR), Institute of Transportation Systems, Germany

ABSTRACT

Currently, there are only a few technical aids or even control methods to accelerate autonomous buses in urban traffic or to provide them with an extended green phase. This is necessary because their speed is very different from other road users. Communication between autonomous buses and the infrastructure using vehicle-to-everything (V2X) communication is already possible and well researched. It is not yet supported by many manufacturers of the control units and also manufacturers of vehicles. In the project, we have researched the development of a new traffic light signal system (LSA) control procedure with innovative traffic light system communication and detection technology and also implemented it in real traffic. The results from the research can currently already be used for small vessel sizes. The offer can be extended by possible on-demand stops and the correct reaction to vehicles with official duties like police, fire department or ambulance vehicles. In particular, interchanges without a driver present or the usual stop infrastructure at virtual on-demand stops require more passenger guidance by autonomous transport modes.

Keywords: V2X, C2X, Car2x, VITAL, shuttle, autonomous, traffic light controller, 802.11p.

1 INTRODUCTION

The networking of vehicles via V2X [1] interfaces also enables participation in cooperative driving applications. Cooperative behaviour in the sense of individual actions of single vehicles based on jointly collected data has been demonstrated and investigated, for example, in the publicly funded project Transaid [2] and Maven [3].

In the BMVI-funded project "SIRENE" [4], the networking of special forces with intelligent infrastructure was investigated. The aim was to enable optimal routing, considering the current traffic situation. Intelligent traffic lights communicate with the emergency vehicles, priorities the emergency vehicles in case of an emergency and thus enable a green wave for them. The communication took place via 802.11p (V2X).

The networking of vehicles was also investigated in the research project "KoMoD" [5]. At intersections, public transport traffic was prioritized by means of phase shifting or green time extension. For this purpose, information about the position, direction of travel, speed and line and route are transmitted on the vehicle side.

The acceleration of public transport vehicles has been known for a long time [6]. In order for the acceleration of public transport at traffic lights to be successful, additional components are required. An appropriately equipped vehicle and a radio-capable traffic light control unit. In order for the traffic lights to receive the signals and thus give priority to public transport, they must be equipped with radio receivers and an adapted traffic light or signal program. The traffic light systems are controlled by sending data telegrams via analogue radio data Transmission. These are sent to signaling points. These are virtual points on the route that were defined with meter precision in the planning phase. In addition to information about the reporting point, the transmitted "telegrams" also contain the line and course number of the requesting vehicle. The basic principle of a pre-announcement, main announcement and sign-off, as well as reporting points, is to remain in place. The only difference is that communication will now take place with the help of V2X communication and a so-called

www.witpress.com, ISSN 1743-3509 (on-line)
doi:10.2495/UMT220011

VDV standard telegram R09.16 telegram [7] in order to allow autonomous public transport to communicate with other road users and to cover other use cases.

The paper should be used to explain this principle in more detail. First, the V2X communication is explained in more detail. A range measurement was then carried out, the results of which can be found in Section 3. In Section 4, the new traffic light control methods are presented and how they are used. Section 5 explains the technology, how the RSU communicates with the OBU on the bus and what data is exchanged. Finally, a test was carried out on a test site in order to have tested the new process before it is used in real traffic. In the last section, the training and the result are explained.

2 V2X COMMUNICATION

The IEEE 802.11p [8] communication standard has been defined together with a special purely software-defined protocol for information exchange between roads vehicles. Therefore, it is called Vehicle2Vehicle (V2V) it can be used to communicate with the infrastructure as well and is then referred to as Vehicle2X (V2X) communication. Wider use in the automotive environment will lead to high market penetration [9] and availability as well as extremely low cost for the technology. Here, messages will be exchanged between vehicles using car-to-car communication (C2C) or via roadside units (RSU) using car-to-infrastructure communication (C2I).

V2X is a WLAN-like standard used in the automotive sector to exchange information between vehicles and/or the infrastructure [10]. This messaging standard is used, for example, in road traffic to immediately warn and support drivers in emergency situations. For this reason, V2X components have been developed and tested for several years. However, this communication standard is not limited to the automotive industry, nor should it have anything to do with the means of transport. The use of V2X communication in road traffic has been evaluated in a number of projects. In principle, it is possible to exchange information across modes of transport.

For the testing and evaluation, use cases in the field of public transport were defined and designed that enable economic operation through V2X application. On the basis of these use cases, it could be shown that on the one hand the use of V2X standard is possible and on the other hand the use of cost-effective standard components from the automotive sector is also target-oriented in public transport.

In order to enable communication between vehicles and infrastructure, which is relevant, e.g. for intelligent assistance systems in vehicles, standards for vehicle-to-infrastructure communication (V2X) have been developed. The European Telecommunications Standards Institute (ETSI) has developed a set of standards for this purpose.

V2X messages are exchanged between receivers via the extended WLAN standard IEEE 802.11p in the 5 GHz frequency range. Research is also being conducted on V2X communication via mobile networks. Studies show that although LTE achieves higher latency than communication via 802.11p, it has better scalability and enables communication over long distances or worldwide.

For data exchange in the WLAN standard IEEE 802.11p, the ETSI ITS G5 standards [11] specify the required network architecture. The ETSI ITS G5 was developed to describe the architecture and network mechanisms for V2V (vehicle-to-vehicle) and V2I (vehicle-to-infrastructure, e.g. roadside units) communication.

Four standardized messages were used in the research project, which are briefly explained below: Cooperative Awareness Messages (CAM) messages are status information about vehicle position, driving speed, driving direction, vehicle status The CAM messages are sent out once per second. The Decentralized Environmental Notification Message (DENM) is a message about location and traffic data as well as traffic situations and danger spots. The Signal Phase and Timing Extended Message (SPATEM) is real-time information transmitted by control units. The MAPEM contains information about the topology of a road or intersection that is required for automated vehicles – in principle the map of the traffic light system.

3 RANGE MEASUREMENT

Since we have no information about the local conditions in the different test environments, we first need to measure how far the V2X messages can be sent and received. This is done using a range measurement, which has already been tested many times and is therefore accurate enough to provide an initial estimate of the correct transmission of V2X messages. The range measurement was performed with a Cohda Wireless MK5 [12] RSU – as base station, as well as with a Cohda Wireless MK5 OBU in the vehicle – with the latest available firmware version. The antennas used are originally supplied by Cohda. The recording was done with the help of Cohda's own field measurement tool, which is included as standard on every Cohda wireless box. The evaluation of the generated files was done with a DLR tool specially developed for range measurement. The modulation used was a 12QPSK, which calculated the range with different packet lengths of 100, 400, 800, 1500 bytes. The location of the measurement was Berliner Straße in Berlin Reinickendorf, compare with Fig. 1.

The RSU has the following coordinates: 52.584175, 13.288600.

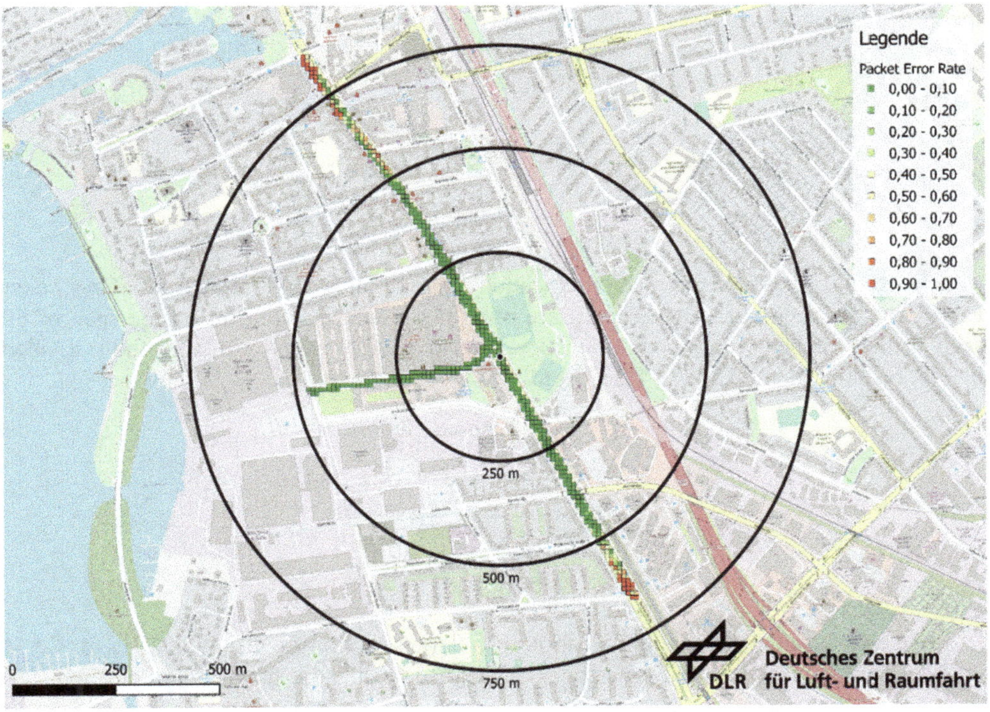

Figure 1: Range measurement with PER.

The evaluation is based on the packet error rate (PER) [13]. This error rate refers to the data packets that were not received. Non-received data packets can be recognized by means of the message ID. During the evaluation, several messages in a previously defined measuring section are combined to form a measuring point. The PER of these measuring points are averaged and designated as PER. Three categories were defined to subdivide the PER at the measuring points. A PER of less than 40% is considered the best possible result and is marked with a green dot on the overview map. A PER between 40% and 60% is still considered an acceptable result, as increasing packet losses can already be expected here. The reception of messages can also no longer be accepted with multiple transmissions. In the overview map, these measurement points are marked with a yellow dot.

The last category is a PER of more than 60%. Here, the reception of a message can almost no longer be accepted. Occasionally, reception may still be possible, but the message may be so faulty that it can no longer be used. These measuring points are marked with a red dot in the overview map.

The subsequent evaluation of the results shows that in urban areas a range of about 500m is achieved due to the coverage by buildings and trees. Since 50 km/h is often the speed limit in German city centre, we calculate with this value for better comprehensibility. The speed limit is given in km/h and is divided by 3.6, resulting in m/s. That is 13.88889m/s.

Formula for uniform motion:

$$\mathrm{s} = \mathrm{v} \times \mathrm{t},$$

where is "s" the distance in meter (m), "v" is the speed in meter per second (m/s) and "t" is the time in seconds (s).

$$500\ \mathrm{m} = \frac{50\ \mathrm{km}}{\mathrm{h}} \times \mathrm{t},$$

$$\mathrm{t} = \frac{500\ \mathrm{m}}{13.88889\ \mathrm{m/s}},$$

$$\mathrm{t} = 36\ \mathrm{s}.$$

This result corresponds to a communication range of 36 seconds at 50 km/h. On smaller access roads, speeds of only 30 km/h are often permitted, resulting in a time span of 60 seconds that can be used for communication. The speed limit is given in km/h and is divided by 3.6, resulting in m/s. That is 8,33333 m/s.

$$500\ \mathrm{m} = \frac{30\ \mathrm{km}}{\mathrm{h}} \times \mathrm{t},$$

$$\mathrm{t} = \frac{500\ \mathrm{m}}{8{,}333333\ \mathrm{m/s}},$$

$$\mathrm{t} = 60\ \mathrm{s}.$$

If we now assume V2X communication with a transmission frequency of 1 Hz for CAM messages and 10 Hz for DENM messages, this already results in stable communication.

4 VITAL PROCEDURES: TRAFFIC-DEPENDENT INTELLIGENT CONTROL OF TRAFFIC LIGHTS [14]

Intelligent control procedures for traffic light systems (LSA) represent an important element for traffic control. Against this background, two novel traffic signal control methods were developed at the German Aerospace Centre, which consider the central technical innovations in the field of traffic data acquisition and thus contribute to an improved traffic management.

The essential innovation of these procedures lies in the traffic parameters used for control and the infrastructure used for their acquisition. Up to now, traffic lights have not been controlled with variables such as lost time, and the vehicles themselves, with their built-in recording and communication options, have not yet been used as sensors for traffic light control.

The experience from the projects and also from the knowledge about the novel traffic light systems control procedure should enable us to integrate the autonomous buses into road traffic in such a way that they can be accelerated better in urban traffic or be provided with an extended green phase.

5 EMPIRICAL ANALYSIS

The first step was to set up the system using the bus registration. Here, a pre-announcement was carried out first. This signals the approach of the public transport vehicle to the intersection in advance. The circulation for the priority of the direction of travel is prepared in the control unit of the traffic light. Subsequently, the registration at the traffic light systems is executed by the public transport vehicle. This indicates the immediate arrival of the public transport vehicle at the intersection. The priority of the direction of travel of the public transport vehicle is initiated in the traffic light systems. If a pre-announced public transport vehicle is at a stop in the intersection area, the announcement is only made when the door is closed and the vehicle starts. Finally, the de-registration is carried out. Here, the control system is informed that the vehicle has passed the intersection. The traffic light systems return to its regular cycle.

In the previous section, the regular procedure of a public transport prioritization was presented. During the research project, however, V2X communication is to be used for the autonomous shuttle buses of the public transport system. Thus, the investigation of a procedure for the integration of V2X-based traffic control information for cooperative applications was aimed at. Here, the decision making has to be placed on V2X messages. For the system, SPATEM/MAPEM messages or CAM/DENM messages were exchanged. The traffic light systems send out the messages with the help of an RSU. These can be received up to 500 m from the intersection, as described in the section above. The public transport vehicle also uses an OBU to send out the V2X messages received from the RSU. A simple illustration of the system is shown in Fig. 2. Here, the control unit is equipped with an RSU to send V2X messages. The public transport vehicle has a C-ITS (Cooperative Intelligent Transport Systems) OBU and sends out Cam or DENM messages, which are received by the RSU at the control unit. Meanwhile, the traffic light system sends out its status and topology information to all road users using the SPATEM and MAPEM.

During the automated navigation, implementation and testing of selected cooperative applications for traffic control using V2X communication, cooperative applications for traffic control using V2X technology were tested as a real-world laboratory test on DLR premises and in the test field [15]. For the tests, corresponding modules were implemented on the test vehicles.

6 SCENARIO TEST AND RESULTS

For the scenario test, an intersection was set up in Berlin. Here, detection hardware and the corresponding control system were implemented in a control unit from the laboratory. Care was taken to ensure that the intersection in the test field was set up exactly like the intersection

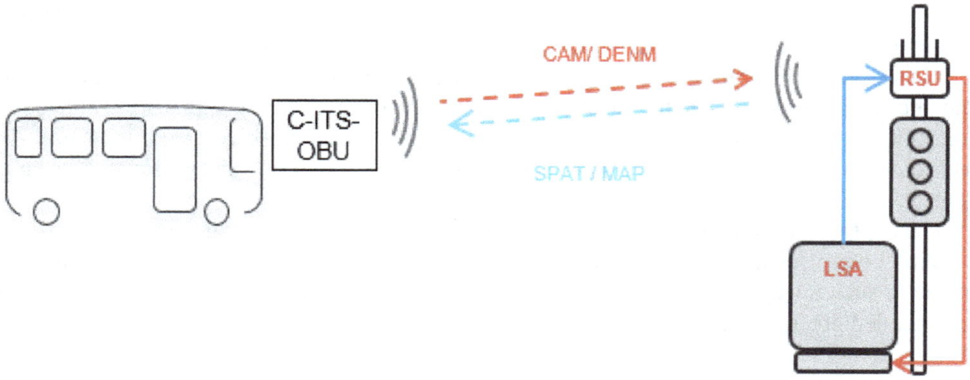

Figure 2: System setup with V2X messages.

in the real world. It has been set up as a four-arm intersection where a main road – in this case Berliner Straße – is crossed by Brunowstraße as shown in Fig. 3. A complete public transport registration was carried out at the control unit as described in the system setup. Subsequently, the sign-off was sent by the test vehicle and the prioritization for the vehicle was cancelled again. The test intersection can be seen in Fig. 3.

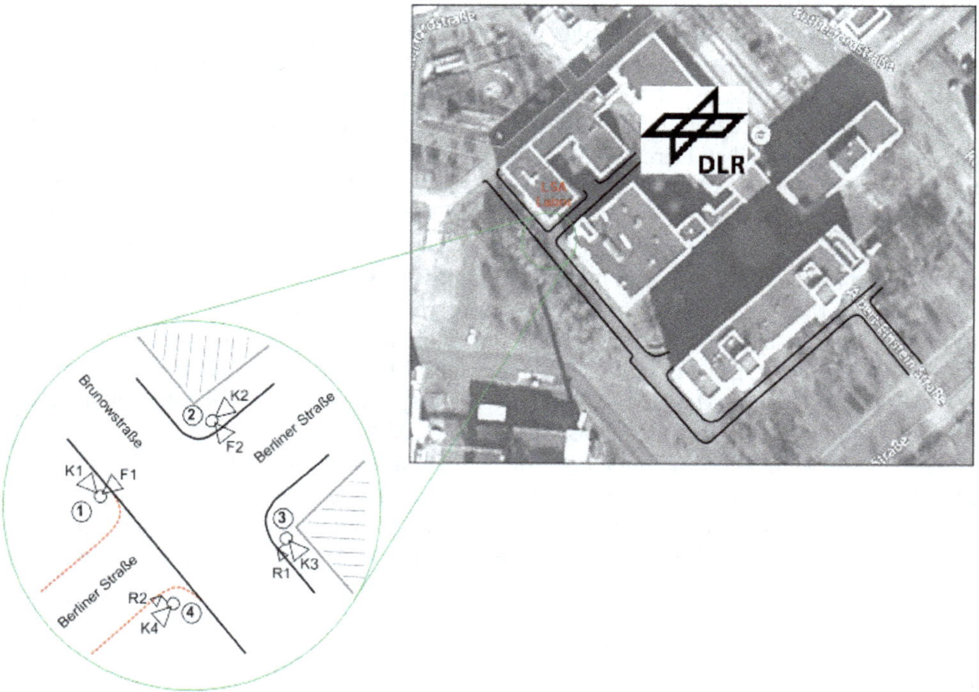

Figure 3: Test intersection on DLR site.

In order to be able to test the detection hardware in the field, it was equipped with detection hardware of the type ThermiCam2 [16] from FLIR at a junction where the range measurement was also carried out. This has an integrated heat sensor and detector for vehicles, vulnerable road users such as cyclists and pedestrians as well as the detection of wrong-way drivers. The ThermiCam2 also works in the dark, as the system only detects the heat emission of road users. Thus, the sensor detects vehicles and vulnerable road users even at night, at long distances and in unfavorable weather conditions. This data is then processed in a simulation and thus feeds back into the VITAL Control system, helping to implement a better traffic flow by changing control procedures. This implementation can be seen in Fig. 4. The C-ITS vehicle communicates with the traffic light systems and the RSU. A long-term test is currently in progress. Currently, only data from April 2022 onwards can be used for analysis. MAPEM and SPATEM messages have also been created for both intersections to allow the autonomous vehicles to communicate with the intersection.

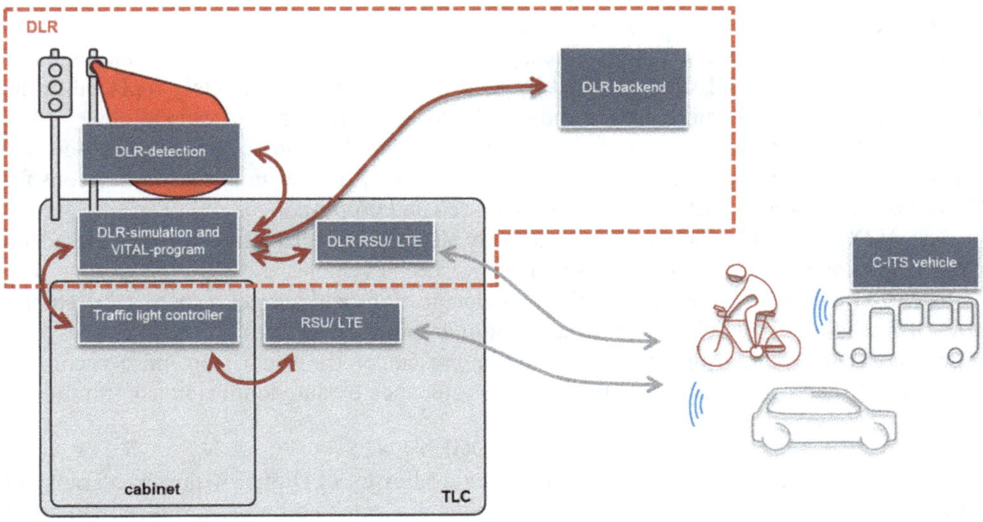

Figure 4: System overview.

7 CONCLUSION

In the context of the Shuttles & Co [17] project, it was shown that V2X communication does not have to be limited exclusively to the automotive sector, but that there are also sufficient fields of action for the public transport sector. During the project, the focus was on the exchange of information between a public transport bus and the infrastructure system in a use case. The research project can also report a further development with regard to the VITAL procedure, in that for the first time the lost times or predicted arrival times at the stop lines of the vehicles will be considered for the control of the traffic light systems. This is possible through the communication of the intelligent traffic lights with the cars using the built-in detection and V2X communication technology. With the lost-time-based and the cooperative method, two novel approaches to traffic light systems control have been developed that are fundamentally based on information from vehicle-infrastructure communication (C2I). These have so far been tested and evaluated in extensive simulation studies. Since vehicle-to-infrastructure communication (C2I) between traffic light systems and vehicles is not yet sufficiently available, an interim solution with fixed detectors from the company FLIR had to be created to record the characteristics. The two new traffic light systems methods were then integrated into the existing control systems in such a way that their structure and basic parameters remained unchanged. The loss-time-based method can be completely mapped on a standardized traffic light system control unit independent of the manufacturer, while the cooperative method requires an additional computing unit for the exchange of information with the traffic light systems control unit. With the future use of vehicle infrastructure communication (C2I), control procedures could change fundamentally, even far below the equipment rates of full equipment. Since the VITAL procedures have so far only used information from private vehicles for control purposes, there is a need for further research and development in the integration of public transport, special purpose vehicles, pedestrians and cyclists. In principle, these can already be integrated into the procedures, as they do not distinguish from which road user, for example, information on position, speed or lost time status originates. However, a separate treatment would be advantageous here – especially for special purpose vehicles with rights of way, not to mention the challenges in detecting vulnerable road users such as pedestrians and cyclists.

The special feature is also that the already existing fixed infrastructure can continue to be used for this purpose and supplemented by V2X technology or completely replaced in the future. This saves infrastructure costs and enables a short-term market introduction.

The V2X communication solution presented here is also subject to a security-critical aspect in terms of availability, integrity and confidentiality. For security reasons, options for the use of public key infrastructures were discussed and proposed.

The V2X communication protocols already defined by ETSI are currently based on communication between road vehicles and between road vehicles and infrastructure.

The offer can be extended by possible on-demand stops and correct response to Vehicles with official duties like police, fire department or ambulance vehicles.

In particular, interchanges without a driver present or the usual stop infrastructure at virtual on-demand stops require more passenger guidance by autonomous transport modes.

ACKNOWLEDGEMENTS

The work presented here was funded by the Federal Ministry of Digital Affairs and Transport (BMDV) in the AVF Projects program. The results presented here were produced as part of the project “Shuttles & Co – Autonomous Shuttles & Co in the digital test field of urban traffic” (FKZ 01MM19011-C).

REFERENCES

[1] Car 2 Car Communication Consortium, V2X Communication. https://www.car-2-car.org/. Accessed on: 10 Mar. 2022.

[2] TransAid. https://www.transaid.eu/. Accessed on: 15 Mar. 2022.

[3] Dariani, R., Schindler, J., Leich, A., Matheis, D. & Rondinone, M., Maven Deliverable 3.1: Detailed concepts for cooperative manoeuvre and trajectory planning and for in-vehicle cooperative environment perception. Projektbericht. 2019. https://elib.dlr.de/133590/. Accessed on: 15 May 2022.

[4] Ruppe, S., Nippold, R. & Trumpold, J., SIRENE: Secure and intelligent road emergency network. *SUMO User Conference 2018*, 14–16 May, Berlin, Germany, 2018. https://elib.dlr.de/123280/. Accessed on: 15 May 2022.

[5] KoMoD. https://www.komod-testfeld.org/. Accessed on: 12 Mar. 2022.

[6] Franco, G. & Biora, F., Public transport priority systems: Opportunities and recommendations. http://www.its.leeds.ac.uk/projects/primavera/deliv_6.html. Accessed on: 15 May 2022.

[7] Schemmel, H., R09.16 Data set for reporting systems. VDV-VÖV Schriften Reihe Technik Nr. 04.05.1/Ergänzung 2 Ausgabe Juni 1990. https://www.vdv.de/voev-04-05-1-erg.pdfx. Accessed on: 29 May 2022.
[8] IEEE 802.11p. https://standards.ieee.org/ieee/802.11p/3953/. Accessed on: 15 May 2022.
[9] Protzmann, R., Radusch, I., Festag, A., Fritzsche, R. & Rehme, M., IV2X integrated consultation vehicle communication. https://www.testfeld-berlin.de/pdfs/iV2X%20 Dokumentation%201.5.pdf. Accessed on: 24 May 2022.
[10] Zimmermann, W. & Schmidgall, R., *Bussysteme in der Fahrzeugtechnik: Protokolle, Standards und Softwarearchitektur*, 4th ed., Vieweg+Teubner, 2010.
[11] ETSI, 5G standards. https://www.etsi.org/technologies/5G. Accessed on: 24 May 2022.
[12] Cohda Wireless, MK5. https://www.cohdawireless.com/solutions/hardware/mk5-rsu/. Accessed on: 24 May 2022.
[13] Packet error rate (PER) measurement description. https://rfmw.em.keysight.com/rfcomms/refdocs/1xevdo/1xevdo_meas_cperror_desc.html. Accessed on: 24 May 2022.
[14] Trumpold, J. & Oertel, R., Einfluss von V2I auf die Verkehrssteuerung. *Vorstellung des VITAL Steuerverfahrens. Workshop Augsburg – Smart City*, 11 Oct., Augsburg, Germany, 2018.
[15] Digital Testfield Stadtverkehr Berlin. https://testfeldstadtverkehr.berlin/. Accessed on: 24 May 2022.
[16] Flir Thermicam 2. https://www.flir.de/products/thermicam2/?vertical=public%20 safety&segment=solutions. Accessed on: 24 May 2022.
[17] Shuttles & Co. https://www.bmvi.de/SharedDocs/DE/Artikel/DG/AVF-projekte/shuttles-und-co.html. Accessed on: 24 May 2022.

BIG DATA IN SMART TRANSPORTATION PLANNING: TOWARDS AN APPLICABLE SMART SOCIO-ECONOMIC TRANSPORTATION PLATFORM IN HISTORICAL MEGA CITIES

REEM BAKRI, ZEYAD EL-SAYAD & ALI F. BAKR
Department of Architecture, Faculty of Engineering, University of Alexandria, Egypt

ABSTRACT

This paper aims to develop the feasibility of transportation big data in order to drive changes that impacts directly both government and citizens' goals and objectives. The relevance is to face and develop transportation challenges in the down-town of historical mega-cities. A smart transportation platform is proposed that can manage huge amounts of big transportation data planning (BTDP) and give guidance to the urban authorities to make their municipalities smarter and digital, and to propose smart transportation building (STB) as part of smart digital city while preserving its historical identity. This paper proposed a developed framework for smart socio-economic indicator through transportation platform for city data harvesting. The results were able to classify major challenges facing users which revolving the socio-economic indicator; according to a surveying questionnaire on the historical down-town of Alexandria City in Egypt as an important historical mega-city; in order to set applicable solutions via using the Smart socio-economic transportation platform (S2TP).

Keywords: big transportation data planning (BTDP), historical mega-cities, smart transportation building (STB), socio-economic indicator, smart socio-economic transportation platform (S2TP).

1 INTRODUCTION

Big data in transportation is one of the most fashionable concepts nowadays, governments try to exploit the huge new amount of available information [1]. One of the most important problems facing transportation planning in mega-cities is the huge amount of non-upgraded data in order to make it more digital and smarter; while preserving its historical identity. Therefore, heading towards using smart transportation platforms and real-time applications will face these challenges while classifying historical identity.

Subsequently this paper determine down town of Alexandria City in Egypt as a case study in order to specific the maturation of socio-economic composite indicator as a basic return on both social and economic bearings, and goes along the government development plan to improve communication between transport systems and developing economy to correspond objectives of the strategic plan for Alexandria year 2032 [2].

2 BIG TRANSPORTATION DATA FRAMEWORK AND METHODS

Through civil registrations the previous decades saw a similar data avalanche as we are witnessing today, partly through bureaucratic and commercial records, through surveys and mapping, and data has subsequently come to underlie city planning and decisions [3].

2.1 Purpose of big data framework

People should evaluate the purposes for which their data is used and weigh the benefits that supplying their data may provide. When the rewards are immediate and personal (social services, commercial gain), most people are willing to provide their information with the entity that requests it [4].

WIT Transactions on The Built Environment, Vol 212, © 2022 WIT Press
www.witpress.com, ISSN 1743-3509 (on-line)
doi:10.2495/UMT220021

Benefits of big data sharing with a broader, societal purpose are less readily acknowledged as a worthwhile trade-off. Concern that their data will be used for purposes other than those for which they were acquired is a complicating element for citizens and consumers assessing the purpose of big data acquisition [5]. Determine what kinds of privacy concerns people as citizens, workers, shoppers, or travellers in smart cities may have as a result of the usage of smart transportation open, huge and linked data. On the basis of existing research on privacy perceptions, it is asserted that these concerns are underpinned by people's perceptions of particular data as personal or impersonal as shown in Fig. 1; their concerns differ depending on the purpose for which data is collected service or surveillance, four areas of concern emerge that range from hardly (impersonal data, service purpose) to extremely high (personal data, surveillance purpose) as shown in Fig. 1.

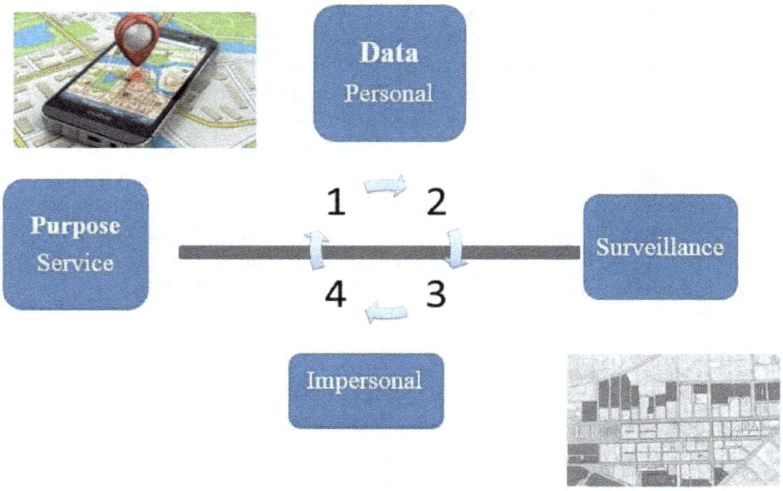

Figure 1: Contrasting challenges for data city transport.

The framework was then used to investigate how specific technology (such as smart parking) and data utilisation (predictive, policing, social interaction) could result in varying levels of privacy concerns. This led to the general idea that smart technology selection and the application of certain combinations of big data and analytic tools are critical variables in understanding people's concerns about smart transportation [6].

2.2 Implementation of big transportation data surveying methods:

Big transportation data has been harvested in this research using three methods:

1. Site visiting real live shots in order to identify historical places challenges
2. Satellite real-time maps as shown on google maps
3. Surveying questionnaire in order to collect data from users (drivers, passengers and pedestrians); which includes visits information, walkability, location identification and usability of smart transportation platforms.

The road real-time maps are placed globally, generating traffic data such as the number of cars and average traffic; it also could be implemented to locate historical places, shortcuts roads, service and car parking availability. The three sources transfer data through the internet to the smart transportation building (STB), which is a component of the smart city building to gather incoming city data over the network and devices via using smart transportation n applicable platform [7].

3 SOCIO-ECONOMIC COMPOSITE INDICATOR IN SMART TRANSPORTATION PLANNING AND IMPACTS

Following the introduction of big transportation data framework and its recommended implementation; thus, it is necessary to determine composite indicator for big data in transportation; which needs to be collected, upgraded and managed in mega cities characterized as historical cities.

Worldwide 55% of the global population growth in mega cities. Thus, the current urban population of 4 billion people is expected to grow to 6 billion by 2050 [8].

Urban planning adopt various forms and characterizations, and urban growth patterns change between regions, as a result of socio-economic, cultural and historical variances. Therefore the speed of transportation presents challenges which need to be settled and solved according to a smart socio-economic indicator as a main composite impacted indicator, essentially in the down town of mega-cities.

3.1 Socio-economic data collecting of mega-cities

Data about socio-economic has been collected from a number of cities around the world [9]. The data utilised is assumed to be comparable across its impact on historical down-town of mega cities allowing for a global development of the transportation performance indicator. The performance composite indicator for this study is the weekly visits for users to the down town district; facilitating the purpose of the visits, whether it is for business, trade, visiting historical places, or for entertainment, shopping and walking.

3.2 Developing socio-economic mainstream data parameters

A smart city links human capital, social capital, and infrastructures to handle public concerns, develop transportation, and improve citizens' quality of life.

Indeed, when considering the success criteria for a smart transportation deployment, it is clear that citizens have precedence over the other enablers, as in the economic strategy scheme, people, process, and technology must engaged, informed, and connected citizens [10]. All components are deployed in a comprehensive and consistent manner, towards a smart citizen including passengers, drivers and pedestrians

Socio-economic parameters including solving and improving citizens' quality of transportation usability, in parallel big infrastructure data management including vehicles to vehicles (V2V) in order to interact with each other and vehicular networks (VN).

3.3 Transportation digitalization for global world economy

The transportation industry is one of the most heavily influenced sectors of the economy by digitization. The application of technologies that have been successfully tested in big data and intellectualization processes to the transportation sector [11].

As a result, smart transportation systems (STS) are the industry's key technological development trend. The digitalization of the transportation business necessitates a shift in the production's core technical and economic basis. There are important aspects of the transportation sector's digitalization effort at the moment; thus transportation infrastructure and supply chains are becoming more digital, including management procedures and automation on a large scale [12].

3.4 Socio-economic impacts

Big data analysis, multi-criteria evaluations, and data envelopment analysis (DEA) are the most prevalent assessment approaches for socio-economic assessments. These methods are used to evaluate the socio-economic consequences of intelligent transportation systems [13]. Economic, social, transportation, and environmental integration, which would ensure a decent existence for at least several generations of citizens [14].

Information, knowledge, widespread use of massive transportation data planning, and information and communication technology all play important socioeconomic roles.

The city is an economic and social laboratory in which the citizen image of the city is generated at each step of its historical evolution [15].

4 TRANSPORTATION IN ALEXANDRIA'S HISTORICAL DOWN TOWN CHALLENGES SURVEYING AND IDENTIFICATION

Alexandria is one of the most important mega cities in the world with a distinct historical and cultural identity. Likewise, it is known as the cosmopolitan city [16], and according to the increase of population explosion in several areas, especially the downtown district, and the number of people visits, especially on rush hours times, and the multiplicity of users and visitors, whether for work, trade, services or entertainment; such as shopping and walking.

As per the researcher's evaluation during previous surveying questionnaire, transportation has become afflicted by a number of challenges. According to the strategic urban plan of Alexandria City 2032; the strategic pyramid defines developments tenors and alternatives, which settle that one of the most effective tenor is to define the identity transportation in Alexandria as a smart city with a brilliant history.

4.1 Transportation challenges in down-town of Alexandria City

Historical down-town in Alexandria City in Egypt includes several important landmarks, land uses and significant characteristics are handled together at same boundaries. Therefore, demanding networked management process are necessity.

Challenges in down-town of Alexandria could be summed up into five major determinants as shown in Fig. 2: disconnected users, walkability and pedestrianism lanes or platforms, public and private disconnected vehicles, big traffic non-managed data including historical places and unavailability of car-through services. Subsequently, after identifying challenges; proposals could be provided.

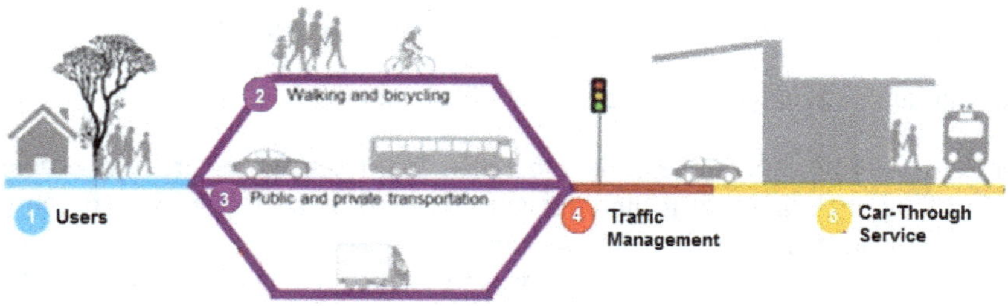

Figure 2: Info-graph shows transportation challenges in down-town of Alexandria City. *(Source: [17], edited by researchers, 2022.)*

4.2 Surveying and identification using satellite maps integrated by real live shots

4.2.1 Historical identity locating and definition

Not all of historical buildings and streets in the down-town of Alexandria City are visible or identifiable on satellite maps as shown in Fig. 3; therefore, big data of historical identity must be updated on the smart transportation real-time platforms.

Figure 3: Historical identity in down-town using real live photos. *(Source: Google maps, and photos by researchers, 2022.)*

4.2.2 Car parking spaces availability

There are numerous of parking spaces that are non-organised and inadequate as they are not upgraded nor connected to a smart network as shown in Fig. 4; thus, many problems occur that cause congestion and traffic jams. On the other hand, using networked vehicles upgraded on real-time smart platform reduces stated problems.

Figure 4: Down-town car parking spaces. *(Source: Google maps, photos by researchers, 2022.)*

4.2.3 Side-roads accessibility suitable for shortcuts

Most of users are accustomed mainly to the use of the main streets and roads; As a result, traffic is completely disrupted during rush hours. Despite the availability of secondary shortcuts that can be used as another alternative to reduce traffic density and solve many road disruption crises as shown in Fig. 5, these shortcuts are not described or updated on smart transportation platforms.

Figure 5: Side-roads in down-town of Alexandria City, Egypt.

4.2.4 Car-through services applicable lanes

Alexandria's downtown district is one of the city's most important commercial centres, among many services such as shops, national service centres, and banks; as a result, users need to perform many services while driving cars to save time and avoid looking for car parking places for long periods of time. Most of service centres and banks are located on the city's major roadways as shown in Fig. 6, allowing the development of a car-through services lanes to reduce wasted time and overcrowding; assuming of necessity, all these lanes and services consistently have to be upgraded on smart transportation platforms.

Figure 6: Roads for car-through service lanes proposals.

4.2.5 Walkability and platforms

Creating developed crosswalk pathways is one of the most important elements in the downtown district of Alexandria City in order to illustrate and identify the historical identity. Thus, to increase the number of visitors and tourists to the historical down town of mega-cities, to get acquainted with it, whether for shopping or entertainment. Even satellite maps lack this important determinant socio-economic indicator. Hence the importance of upgrading smart transportation real-time platforms and providing them with crosswalk pathways for pedestrians as proposed by researchers on satellite map in Fig. 7.

Figure 7: Crosswalk pathways proposed by researchers in down-town of Alexandria, Egypt. *(Source: Google maps, photos by researchers, 2022.)*

5 SOCIO-ECONOMIC INDICATOR SURVEYING ANALYSIS IN DOWN-TOWN OF ALEXANDRIA CITY, EGYPT

Due to applying a surveying questionnaire on 50 citizens through 20 questions by the researcher in the historical downtown district in Alexandria City; citizens selected by different categories in terms of age, gender and classification between passengers and drivers of private vehicles. The researcher was able to settle main problems to ask users about.

5.1 Identifying citizens' categories and their problem statements results

The common problems which citizens face by transportation system as shown in Fig. 8, car parking availability, feasibility of drive-through car service, The usability of short-cuts during rush hours, their Satisfaction of available smart real-time transportation platforms, percentage of their personal usage of real-time transportation platforms, level of satisfaction about walkability and public spaces, identifying historical architecture in buildings and side roads, reasons and numbers of visits to the down-town district, and finally most important proposals of down-town development which fit your needs.

5.2 Citizens' considerations of transportation in down-town development

After applying the surveying questionnaire and rearranging results data into charts to classify them the researcher can come to a conclusion about citizen's observation of the historical

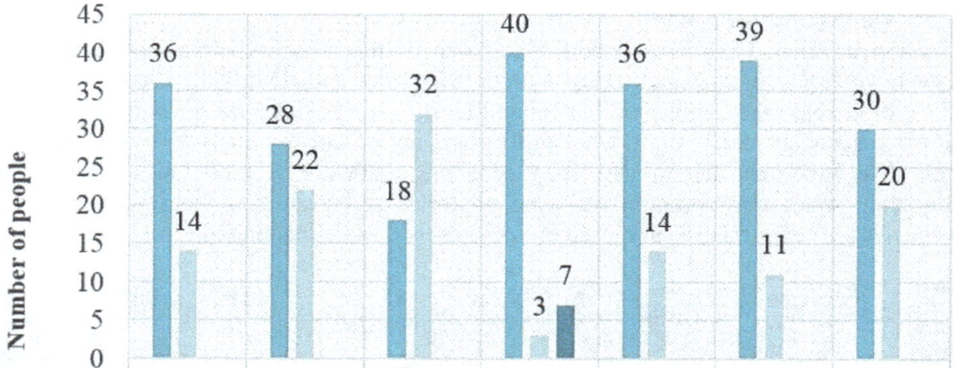

	Gender(Male/Female)	Category (Passenger/Driver)	Own private vehicle/public transportation	Problem (Rush hour/Accessibility/Other)	Car parking availability (Yes/No)	Drive through service (Yes/No)	Shortcuts (Yes/No)
Column1	36	28	18	40	36	39	30
Column 2	14	22	32	3	14	11	20
Column 3				7			

Figure 8: Diagram chart shows classification of citizens and their problems statement about transportation in down-town.

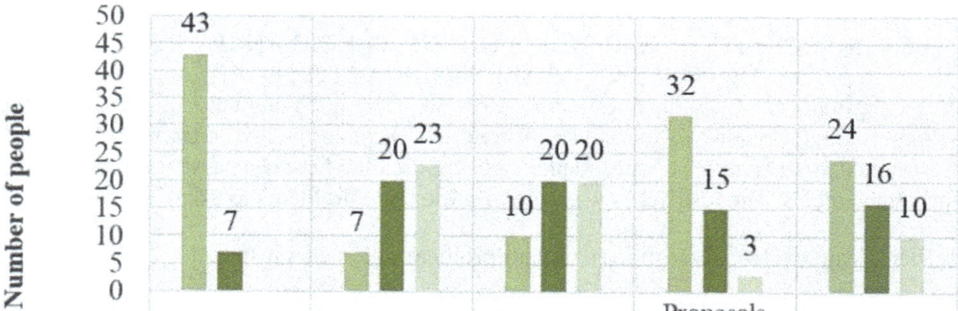

	Architectural Identity(Noticed/ Nothing)	Classification (Historical/Commercial/Both)	Visiting Reasons (Home/Work /Ent.)	Proposals (Trans. dev./ Car parkings/Other)	Benefits (Time/ Comfort/Environment)
Column 1	43	7	10	32	24
Column 2	7	20	20	15	16
Column 3		23	20	3	10

Figure 9: Diagram chart classify citizens' visits, identification and considerations of down-town.

identity of architecture in the down-town of Alexandria City, reasons of citizens' orientations to the down-town; and number of visits per week. The most important valuable benefits of developing transportation in the district of down-town which meet the needs of citizens have also been classified, all results data processed into diagram charts as shown in Fig. 9.

5.3 Citizens' percentage of satisfaction about smart transportation platforms

The researcher can observe citizens' satisfaction about the role of current smart transportation platforms to solve out their need in order to face complicated challenges in transportations in down-town of Alexandria City; through collecting data and processing into diagram chart shown as shown in Fig. 10. Although, results could clarify challenges of:

- Usability of current transportation real-time applications and platforms.
- Satisfaction of walkability and bikes service feasibility
- Historical buildings identification.
- Non-vehicles period proposal acceptance.

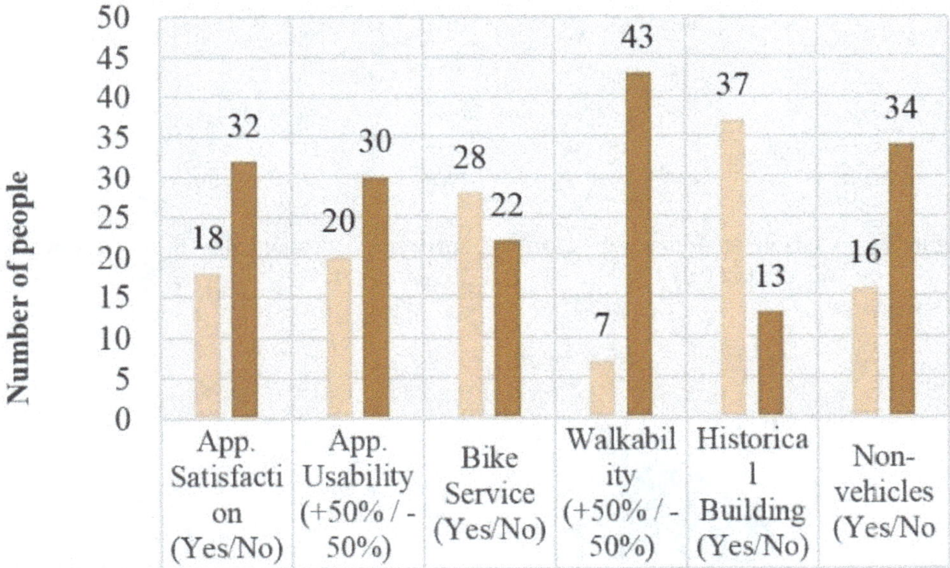

	App. Satisfaction (Yes/No)	App. Usability (+50% / - 50%)	Bike Service (Yes/No)	Walkability (+50% / - 50%)	Historical Building (Yes/No)	Non-vehicles (Yes/No
Column 1	18	20	28	7	37	16
Column 2	32	30	22	43	13	34

Figure 10: Diagram chart illustrate citizens' percentage of satisfaction about usability of current smart transportation platforms.

5.4 Major challenges in transportation according to consolidated surveying results data

Based on previous diagram charts of surveying results; the researcher consolidates a combined data results into a diagram chart as shown in Fig. 11, and conclusion pie-chart shows percentage of five major transportation challenges in down-town of Alexandria City in Egypt as shown in Fig. 12.

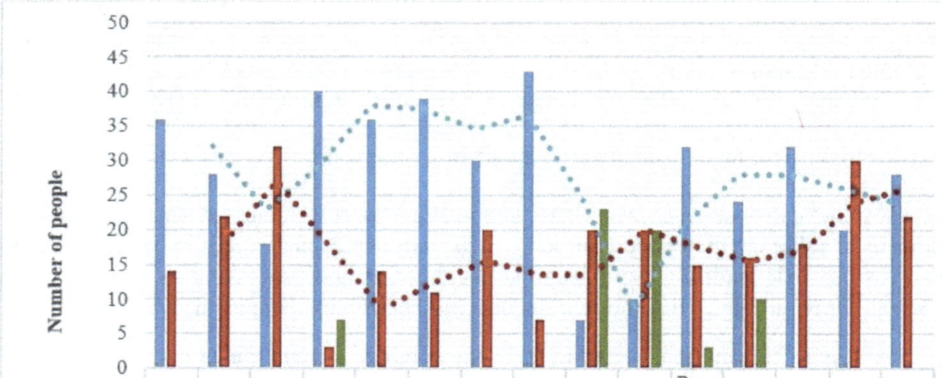

	Gender(Male/Female)	Category (Passenger/Driver)	Own private vehicle/public transportation	Problem (Rush hour/Accessibility/Other)	Car parking availability (Yes/No)	Drive through service (Yes/No)	Shortcuts (Yes/No)	Architectural Identity(Noticed/Nothing)	Classification (Historical/Commercial/Both)	Visiting Reasons (Home/Work/Ent.)	Proposals (Trans. dev./ Car parkings/ Other)	Benefits (Time/ Comfort/Environment)	App. Satisfaction (Yes/No)	App. Usability (+50 % / -50%)	Bike Service (Yes/No)
Column1	36	28	18	40	36	39	30	43	7	10	32	24	32	20	28
Column 2	14	22	32	3	14	11	20	7	20	20	15	16	18	30	22
Column 3				7					23	20	3	10			

Figure 11: Consolidated diagram chart of surveying questionnaire in down-town of Alexandria City.

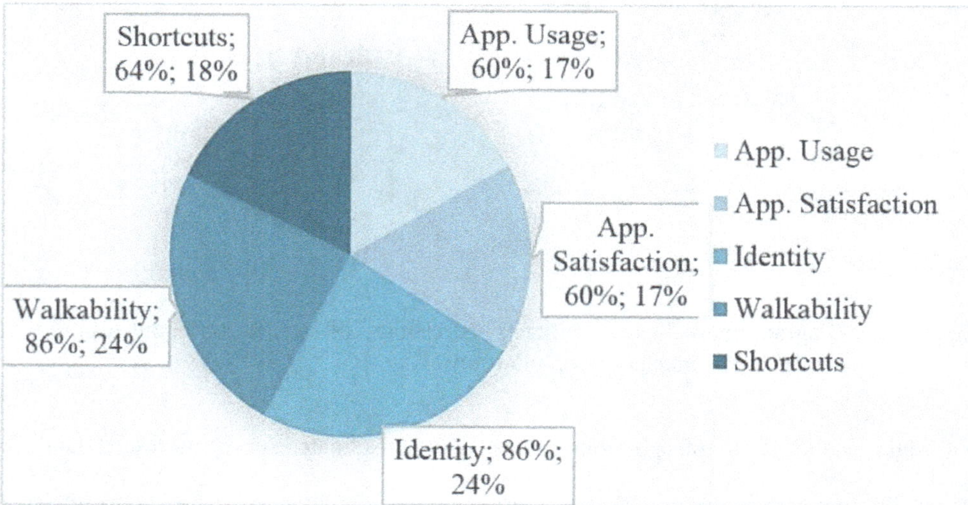

Figure 12: Pie-chart diagram shows percentages of five major transportation challenges in down-town of Alexandria City.

According to the survey combined results chart, the researcher was able to classify five major challenges facing citizens about transportation in 2022, which revolving the socio-economic indicator:

1. Architectural historical buildings identity.
2. Citizens' comfort of walkability and platforms.
3. Shortcuts advantages and accessibility.
4. Smart real-time transportation applications usage.
5. Satisfaction of smart transportation platforms

Following the completion of both analysis and surveying; the results revealed that taking into consideration the socio-economic indicator upgrading is sufficient to accomplish the following outcomes: Transportation: data transportation planning management with networked vehicles. Services: Fully automated services including information, banks and car-through lanes in order to be more engaged and saving time. Integration: A smart city's common information space, integrating data from mega cities infrastructure, big data management systems. Government: Decision-making support system, analysis, incident management, electronic delivery of public and municipal services, open data publication and historical identification. Citizens and users: improving information services, as well as information distributors in historical down-town in mega-cities.

6 CONCLUSIONS:

This paper presents a survey of three methods of big data harvesting from Socio-economic composite indicator in historical mega-cities locating down-town of Alexandria City in Egypt as a case study, and how those surveying results are proposed to be used in an applicable smart transportation platform. The aim of Smart Socio-economic Transportation Platform (S2TP) is to contribute to enhance the transportation planning process through improving big data collecting and management methods for users. (S2TP)'s aim is to upgrade survey data harvesting with network and users' needs inferred from the use of real-time transportation platforms.

REFERENCES

[1] Mayer, V., Schonberger, K. & Cukier, K., *Big Data: A Revolution That Will Transform How We Live, Work, and Think,* Houghton Mifflin Harcourt: New York, 2013.

[2] Strategic Transportation Planning for Alexandria City, Recommended Solutions. Alexandria, Egypt: Strategic Urban Plan Year 2032, 2020.

[3] Robertson, H. & Travaglia, J., Big data problems we face today can be traced to the social ordering practices of the 19th century. *Big Data | Data Science. Politics of Data Series*, 13 Oct., 2015.

[4] Acquisti, A., Leslie, K.J. & Loewenstein, G., What is privacy worth? *Journal of Legal Studies*, **42**, 2013.

[5] European Banking Authority, EBA report on big data and advanced analytics, 2020. eba.europa.eu.

[6] van Zoonen, L., Privacy concerns in smart cities. **33**, pp. 472–480, 2016.

[7] Fusco, F. & Deri, L., High speed network traffic analysis with commodity multi-core systems. *10th ACM SIGCOMM Conference on Internet Measurement*, Nov., Melbourne, Australia, 2010.

[8] The World Bank, Urban development. https://www.worldbank.org/en/topic/urbandevelopment/overview#1. Accessed on: 27 May 2022.

[9] Dingil, A.E., Schweizer, J., Rupi, F. & Stasiskiene, Z., Transport indicator analysis and comparison of 151 urban areas. *European Transport Research Review*, **10**, p. 58, 2018.
[10] Manda, M.I. & Backhouse, J., Smart governance for inclusive socio-economic: Transformation in South Africa. *E-Participation in Smart Cities*, **34**, pp. 179–201, 2019.
[11] Mashkina, N.A., Belyaeva, E.S., Obukhova, A.S. & Belyaeva, O.V., Digitalization of the transport industry in the context of globalization of the world economy. *SHS Web of Conferences*, **92**, 05020, 2021.
[12] Tugashev, A.A., Economics and logistics in the digitalization of the transport industry. **2**, pp. 44–53, 2018.
[13] Zhicai, J., Jianping, W. & McDonald, M., Socio-economic impact assessment of intelligent transport systems. *Tsinghua Science and Technology*, **11**, pp. 339–350, 2006.
[14] United Nations, Transforming our world: The 2030 agenda for sustainable development A/RES/70/1. Department of Economic and Social Affairs, Sustainable Development, United Nations: New York, 2015.
[15] Shichiyakh, R.A., Klyuchnikov, D.A., Balashova, S.P. & Novoselov, S.N., Smart city as the basic construct of the socio-economic development of territories. *International Journal of Economics and Financial*, **6**, pp. 157–162, 2016.
[16] Reimer, M.J., Rowlatt, M. & Mackie, J.A., Alexandria-Egypt. Alternate titles: Al-Iskandariyyah. britannica.com/. https://www.britannica.com/place/Alexandria-Egypt. Accessed on: 7 May 2022.
[17] Prabhakar, V., Gupta, S. & Mehrotra, R., Smart transportation for smart cities. *The Economic Times (ET)*, Sep. 2015.

ALGORITHM FOR PLANNING FASTER ROUTES IN URBAN NETWORKS WITH TIME-DEPENDENT ARCS AND THE POSSIBILITY OF INTRODUCING WAITING PERIODS AT NODES

FRANCISCO A. ORTEGA, GUIDO MARSEGLIA, JUAN A. MESA & RAMÓN PIEDRA-DE-LA-CUADRA
Mathematics Institute (IMUS), Universidad de Sevilla, Spain

ABSTRACT

Navigation systems implemented in mobile devices allow users to search for the shortest routes between pairs of points. Many of the existing commercial products assume in a simplified way that the travel time to cross each arc of a road network is fixed, once a starting time has been established. However, the real travel time along a road section within cities depends on many factors that are related to traffic congestion, weather conditions, possible incidents, etc., and consequently, it depends on the time. As can easily be shown, determining the shortest itineraries in a network whose arcs are time-dependent can result in a diversity of optimal routes for a same origin–destination pair based on different departure times. Assuming the availability of the estimated data of the time required to travel along each section of the street network, once the departure time has been previously set, we propose in this work an efficient algorithm for obtaining faster routes on time-dependent arcs, in such a way that the sum of driving times is minimized, which in parallel allows improving fuel consumption and reducing associated polluting emissions. The possibility of introducing waiting periods in the nodes to optimize the total time spent on the trip has also been considered in the design of the proposed procedure. An experimental evaluation is carried out to show the effectiveness of the provided algorithm.

Keywords: route planning, guided transport systems, mobility, transport networks, Dijkstra's algorithm.

1 INTRODUCTION

From a theoretical point of view, the problem of finding a shortest path from one node to another in a graph with fixed lengths (or fixed travel times) on its arcs is satisfactorily solved. In route planning problems for a single objective, Dijkstra's algorithm (Dijkstra [1]) is preferably used to obtain the solution to the problem of determining the shortest path between two nodes of a connected graph. Its algorithmic complexity is $O(m + n \log n)$, where n is the number of nodes and m the number of arcs of the underlying graph (Cormen et al. [2]).

However, this algorithm in its simplest form can be impractical in many real application scenarios. One of such situations arises when the travel time between arcs is depending on the traffic intensity existing between their respective starting and ending nodes at certain period, which is known as time-dependent arcs. This circumstance can occur both in the determination of optimal public transport routes (bus, metro or dense networks of commuter train) and in private transport routes (own motorized vehicle, bicycle or similar). Therefore, it arouses great practical interest.

As illustration, Fig. 1 shows the evolution of the average intensity in the city of Seville (Spain) throughout the current day (red) compared to the evolution of yesterday (blue), expressed in number of vehicles per lane every 5 minutes. This figure also shows the occupancy levels (green), both today and yesterday, which empirically shows the existence of a correlation between both variables and explains the daily evolution of the circulating difficulty.

WIT Transactions on The Built Environment, Vol 212, © 2022 WIT Press
www.witpress.com, ISSN 1743-3509 (on-line)
doi:10.2495/UMT220031

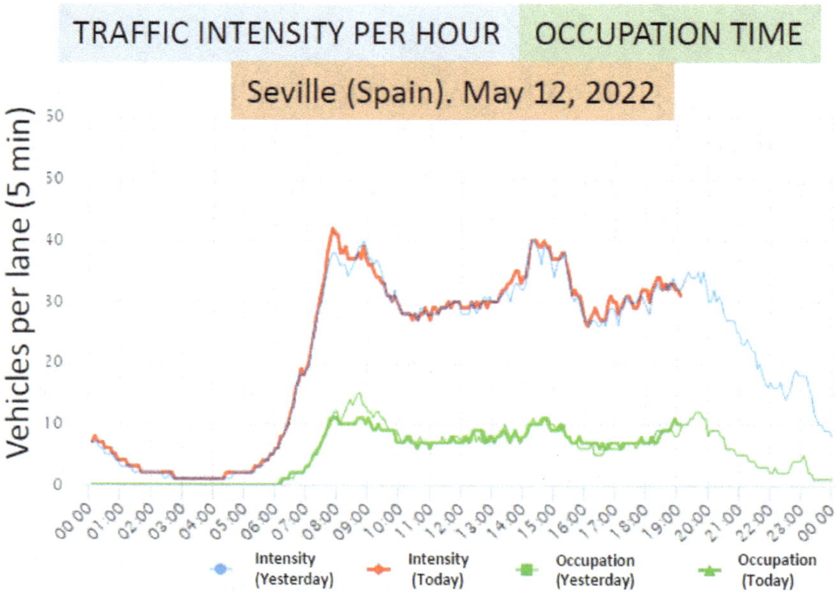

Figure 1: Evolution of the average intensity in the city of Seville along a day. *(Source: http://trafico.sevilla.org/.)*

In general, routing is the process of selecting the "best" paths in a graph G = (V, A), where V is a set of nodes and A is a set of directed arcs. Most studies on routing problems have been carried out under the assumption that all the information needed to formulate such routing problems is time invariant (Toth and Vigo [3]). In many practical applications, this assumption is generally not satisfied since travel times can vary exogenously due to traffic congestion, weather conditions, etc., or endogenously, depending on the decisions freely adopted by the driver, like modifying the speed of the vehicle at its discretion (for example, to adjust fuel consumption) or altering the travel time by including rest periods in driving. A classification of time-dependent routing problems with respect to various criteria can be seen in Pillac et al. [4].

Planning routes for vehicles in general on geographical maps, at a city, region or country scale, is a problem of great practical interest (Preuss and Syrbe [5]). The route calculation offered by Google Maps allows users to know an estimated time of arrival at the selected destination along the itinerary. We can assume that this calculation is made using the maximum allowed speed (1) and the recommended speed in each section (2) as input data. These values, in turn, could be averaged considering the intensity of traffic in real time (3), as well as historical data recorded by other users on previous occasions, if available (4). For a central server that had to respond to a potentially very large number of online requests from clients through a web interface, it would be desirable that the maximum response time of the system, providing a good solution, be limited in a way reasonable (Delling and Wagner [6]). In any case, the procedure followed to obtain these predictions offered by commercial web applications is subject to business confidentiality.

Currently, there are multiple operators that provide information on traffic conditions based on the observed speed of movement of vehicles along the sections. However, this information is not complete since it does not cover all circulating vehicles, either because there is a "non-connected" portion in the fleet of vehicles, or because the sources of this information do not cover all browsers, telephone operators or types of vehicles. For example, Google-Traffic and InfoTransit are two real-time traffic information servers that operate in Spain based on data from telephone operators, specific vehicle fleets (insurers, carriers, automobile clubs, etc.), fixed traffic sensors, fixed traffic and GPS navigation systems. Eglese et al. [7] examine the problems related to the construction of a database of time-dependent travel times for an application developed in England.

Time-dependent routing problems have been treated as a variant of the vehicle routing problem (VRP) for a multiplicity of contexts. Among them (see Gendreau et al. [8]):

- Route planning aircraft, ships or submarines in two-dimensional or three-dimensional space, where decisions include not only the determination of the itinerary, but also the adjustment of power. The time dependency factor can be caused by air currents, ocean swell, or underwater flow. The objective to be minimized is usually the journey time, fuel consumption (or, equivalently, CO_2 emissions) or a combination of them (Perakis and Papadakis [9] and Norstad et al. [10]).
- Other causes that may imply a dependency on the travel time for the determination of the length of an optimal route would be the need to refuel mobile units (Helvig et al. [11]), or the necessary interception of the trajectories of other vehicles (Jiang et al. [12]).
- Finally, the presence of moving obstacles could also be a determining factor when planning the robot movement for such contexts (Sutner and Maass [13], Latombe [14] and Fujimura [15]).

We will assume in this work that the objective that the user always pursues is the achievement of a minimum travel time, which means reaching the destination in the shortest possible time once the starting time for the trip has been established at the origin point. This specific perspective is not the only one that can arouse interest in the specialized literature on transport. For example, public transport users may show different personal preferences when establishing their own concept of optimality (user point of view), which would imply an adaptation of the initial algorithms to accommodate the formulation of other types of optimality. advantageous situations or penalties. Among these adaptations is the consideration of the use of algorithms that produce K-shortest paths (known as KSP problem) to obtain a reasonable number (K) of shortest feasible paths and rank them based on the incorporation of user preferences.

The preferences of public transport users are usually several. As an example:

1. Achieve a minimum travel time, which means reaching the destination in the shortest possible time from the departure time established as the start from the point of origin.
2. Minimize the number of transfers: some travellers prefer to travel in a single vehicle (bus or train), rather than endure the inconvenience of transfers, even though the overall journey time may be longer.
3. Achieving a minimum travel distance route: A traveller loaded with heavy or uncomfortable objects might prefer to walk to the nearest bus stop, as a first step, instead of traveling a greater distance to another bus stop, although this second option would mean a faster route in time.

The calculation of the fastest routes between points in very extensive road networks, whose arcs depend on time, must be carried out through the participation of centralized information services based on an accessible web, where both congestion patterns and traffic data updated in real time are managed.

2 MODELLING THE PROBLEM

Following on, the notation previously used in the work of Dreyfus [16] is now described, where the travel times required to connect a pair of adjacent nodes may be dependent on the time at which the effective travel through the arc begins.

Let the graph $G = (V, A)$ where V is a finite set of n nodes or vertices, and A is a finite set of m arcs that connect the nodes. Each arc can be denoted as the ordered pair (i, j) when it corresponds to the pair of vertices i and j. Let O (origin or initial node) and D (destination or terminal vertex) be two given nodes of G; a path p from O to D in G is defined as the alternating sequence of vertices and arcs: $p = \{O = v_0, a_1, v_1, a_2, v_2, \dots, a_k, v_k = D\}$, such that:

- $a_i \in A, \forall i = 1, \dots k;\ v_i \in V,\ \forall i = 1, \dots k - 1.$
- $a_i = (v_{i-1}, v_i) \in A, \forall i = 1, \dots k.$
- $O, D \in V \backslash \{v_1, v_2, \dots, v_{k-1}\}.$

The cost associated with arc (i, j) is defined as a positive real number c_{ij}. In this way, the cost of a path p will be equal to the accumulated sum of the costs of the arcs that compose it:

$$c(p) = \sum_{(i,j) \in p} c_{ij}. \tag{1}$$

Let P_{ij} be the set of all paths from i to j in the graph G. Following the notation of Dreyfus [16], the problem of finding the fastest path between points O and D where the travel time between point i and point j depends on the starting time of point j can be formulated as follows. Let us denote by $f_i(t)$ the minimum invested travel time until reaching destination D starting from point i at time t. Moreover, let $d_{ij}(t)$ be a positive value that represents the travel time invested when travelling arc (i, j) starting from point i at time t.

The recurring scheme:

$$\begin{cases} f_i(t) = \min\limits_{j \neq i} \ [d_{ij}(t) + f_j(t + d_{ij}(t))] \\ \qquad\qquad f_D(t) = 0, \end{cases} \tag{2}$$

allows us to design an iterative algorithm that, based on the procedure devised by Dijkstra [1], provides the fastest route, in this context of travel times in time-dependent arcs, to obtain $f_O(t)$.

Originally, Dijkstra's algorithm is designed to solve the problem of obtaining optimal paths in networks with not time-dependent arcs. An encoding of this algorithm is presented below.

Dijkstra's algorithm (not time-dependent arcs)

1. **Read** graph **G= (V, A),** adjacency matrix **Ady** and matrix **D** of displacement costs between pairs of adjacent points.
2. **Build** vector **f** that will store the minimum displacement costs to the origin from each node of **G** (initially, only the nodes adjacent to the origin will not have infinite value).
3. **Build** vector **p** that stores the successor nodes to each node of G in the optimal path (initially, only nodes adjacent to the origin can be determined).
4. **Initialize** set **S** of nodes explored by including only the origin point. **Initialize** set **LIST=G\S.**
5. **While** LIST is not empty:
   - **5.1 Identify** index **j***

     $j^* := Arg[\min\{f(j) : j \in LIST\}]$

5.2 Remove index j* from LIST

5.3 For each successor **k** of j* included in LIST:

If $f(k) > f(j^*) + d_{j^*k}(0)$ **then**

a. **Update** $f(k) := f(j^*) + d_{j^*k}(0)$

b. **Update** $p(k) := j^*$

6. **End.**

The following two figures graphically reproduce the same example used in Wen et al. [17] to illustrate the operation of Dijkstra's algorithm and its limitations when the network is time dependent. In Fig. 2 a static situation is considered, where the weights of the arcs do not vary with time.

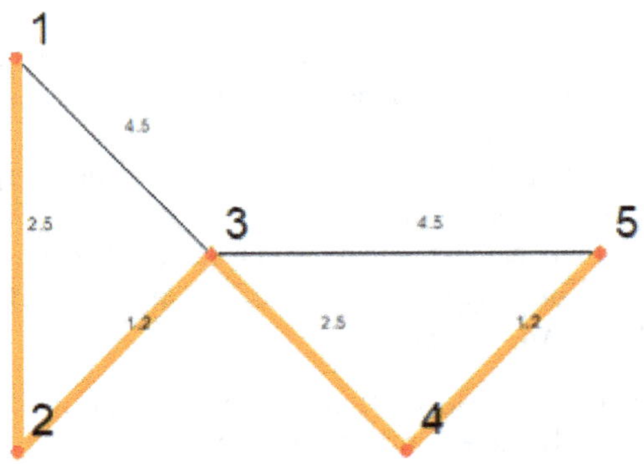

Figure 2: Quickest path between nodes 1 and 5 using Dijkstra's algorithm.

As illustrated in Fig. 2, the solution to the problem is the path 1-2-3-4-5, and the time spent is 7.4 (2.5 + 1.2 + 2.5 + 1.2) minutes, as can be verified by applying the step-by-step algorithm. Let now assume that arc (3,5) is time-dependent in the following way:

1. When the trip along arc (3,5) is started at node 3 in a time t included in the interval [0,4], the total time needed to traverse arc (3,5) is 4.5 minutes, as in the previous static case.
2. However, when the trip through the arc (3,5) starts after time 4, then the travel time needed to traverse arc (3,5) decreases to the value of 1.3 minutes. In that case, as shown in Fig. 3, the fastest path between nodes 1 and 5 would be the sequence 1-3-5, with the time invested corresponding to 5.8 (4.5 + 1.3) minutes.

As found in Wen et al. [17], Dijkstra's algorithm, coded as described above, cannot identify the shortest paths for this context of networks with arcs whose lengths may depend on time. Dijkstra's algorithm is also not initially designed to manage other user strategies, such as setting wait times on some nodes to benefit from shorter travel times after waiting.

Among the descriptive data of this new network context is the identification of those arcs (i,j) whose travel time $d_{ij}(t)$ changes according to the exit time of node i. We will assume that the form of variation of $d_{ij}(t)$ corresponds to a step function that changes its value at certain instants (increasing or decreasing) and remains at that level until the next milestone.

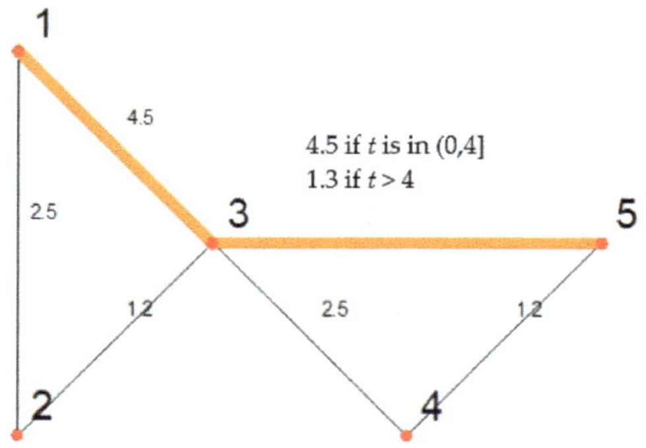

Figure 3: Quickest path between nodes 1 and 5 with time-dependent arcs and no waiting at nodes.

In order to adapt the algorithm of Dijkstra to solve the problem of determining shortest paths in networks with arc lengths that depend on time when starting the trip along the arc, we will make a series of modifications to the original network.

First, we will immerse the planar network in three-dimensional space by raising an axis orthogonal to the base plane for each of the nodes of the network. These axes will indicate the temporal progress at their respective node. For each arc that, starting from node i (at height 0), changes the value of $d_{ij}(t)$ in time t_k, such node i will have to be replicated along its axis, exactly at height t_k. Additionally,

- The distance between nodes i (level 0) and i (level t_k) should be interpreted as waiting time, since geographically there is no change in position.
- The arc (i, j) starting at level 0 indicates that no wait has been carried out at node i by the user before continuing the journey to node j. On the other hand, the arc (i, j) that starts at level t_k indicates that the user has made a wait in node i whose duration is t_k.
- According to the previous assertion, the arc (i, j) that starts at level 0 will be weighted with the value $d_{ij}(0)$, while the arc (i, j) that starts at level t_k will have the value $d_{ij}(t_k)$ as the travel cost.

Furthermore, among the descriptive data of this new network context is the functional characterization of those arcs (i, j) whose travel time $d_{ij}(t)$ changes according to the exit time of node i. We will assume that the form of variation of $d_{ij}(t)$ corresponds to that of a step function that changes its value at certain instants (increasing or decreasing) and staying at that level until the next milestone.

In Fig. 4 has graphically been represented an example of a step function that indicates that the time required to traverse the arc (i, j) is 5, if the start of the trip takes place in time interval [0, 7]; 9, if the start of the trip takes place in time [7, 10]; and so, on until the daily interval [0, 24] is completed.

Let us note that the number of times it would be necessary to replicate a node i along its vertical axis of waiting time will be a function of the number of arcs that start from said node i towards other vertices j and the number of staggered segments $n_i(j)$ that have been considered for explain the behaviour of the function $d_{ij}(t)$. Denoting by *Succ(i)* to the set of vertices j accessible from node i, the number of replications of node i will be, in principle: $\sum_{j \in Succ(i)} n_i(j)$.

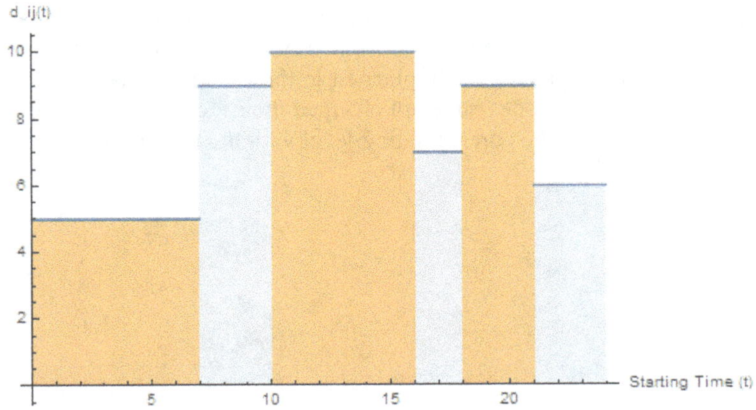

Figure 4: Instance of step function describing $d_{ij}(t)$.

Actually, if we carry out a control of the time progress in each of the vertical axes that start from each node of the network, the number of replicated nodes can be substantially reduced. In fact, it will only make sense to consider the incorporation of the node associated with a travel time $d_{ij}(t_w)$,, obtained after a waiting time t_w at node i, if the sum of both times t_w and $d_{ij}(t_w)$, improves, i.e., is less than the set value without the need to enter waits.

The following example, that will be explained step by step, will explain the proposed methodology. Consider the planar network illustrated in Fig. 5 (8 nodes and 13 arcs), which has previously been embedded in three-dimensional (3D) space. Suppose that origin and destination nodes are labelled 1 and 8, respectively. At the midpoint of each arc of the graph, a numerical value appears, in blue, which corresponds to the time needed to traverse such arc when it is not subject to dependence on the period in which the traversal takes place. In this static context, the fastest path between origin and destination points for these weights is the sequence {1,2,5,7,8} with a total length of 12.5.

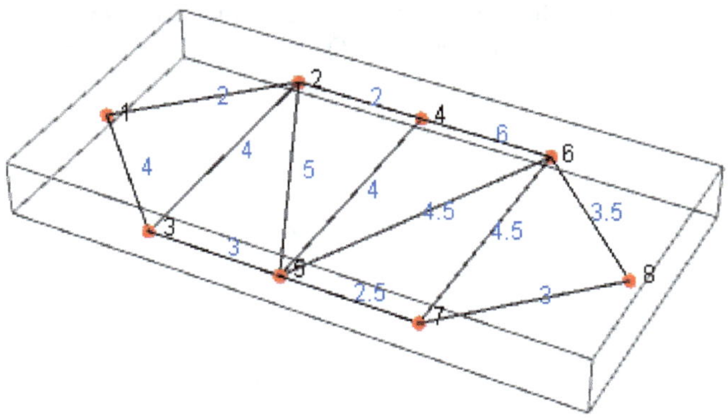

Figure 5: Instance of planar network embedded in 3D-space.

Starting from each node of the network, vertical axes will be established where the new nodes to be replicated can be located. In order to have a reliable control of the time when we analyse the possible alternatives, we will relocate the intermediate nodes (that is, nodes 2–7) to the heights that correspond on their respective axes with the shortest access time from the origin. In Fig. 6, all 8 nodes have been replicated in three-dimensional space (in red). The origin (1) and destination (8) nodes have not changed their location, while the intermediate nodes have been relocated to the corresponding level with the time necessary to reach them from the origin.

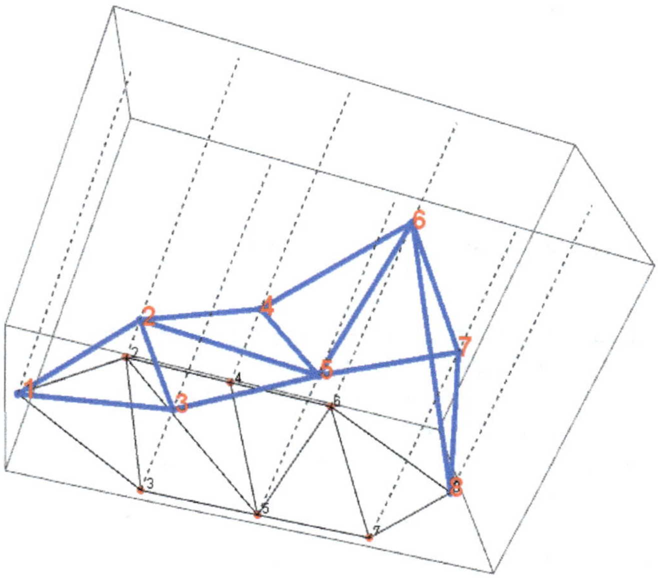

Figure 6: Network with replicated nodes in 3D-space.

Suppose now that two of the three arcs starting from node 4 are time dependent. Specifically, from t = 5, in the arc (4.5) its travel time decreases from 4 to 1.5. For that same mark of t = 5, the arc (4,6) changes its travel time from 6 to 3. Fig. 7 shows the graphs corresponding to the functions $d_{45}(t)$ and $d_{46}(t)$.

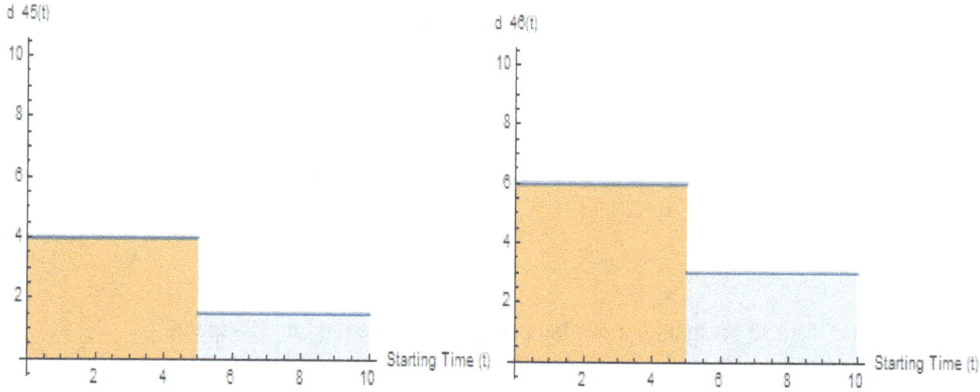

Figure 7: Graphical display of functions $d_{45}(t)$ and $d_{46}(t)$.

Since node 4 is reachable from the origin point (1) in a time t = 4, the user will have two options if node 4 were part of the optimal route to the destination point (8):

1. Immediately explore the arcs starting from node 4 as possible continuation alternatives of an optimal route candidate. This option will not involve modifications in the outgoing arcs from node 4, which graphically is at the elevation level that corresponds to the access time from the origin point.
2. Establish a strategic stop at node 4, with the expectation of obtaining a more competitive travel time on some outgoing arc. This second option forces node 4 to be replicated at the higher vertical level that results after adding the waiting time to the arrival time at the node 4. Node 9, on the same vertical axis of node 4, represents the option to spend one unit of waiting time at node 4 before choosing to continue towards node 5, where the arc (9,5) would be weighted with the value 1.5 instead of 4, or towards node 6, where the arc (9,6) would be weighted with the value 3 instead of 6. Note that it has not been necessary to replicate more times the node 4 because the value changes of the functions $d_{45}(t)$ and $d_{46}(t)$ overlap at the same milestone $t = 4$.

Fig. 8 illustrates the resulting final graph on which, applying a standard Dijkstra's algorithm, the optimal connection route from point 1 to point 8 could be determined, which would be the sequence {1,2,4,9,6,8} with a total length of 11.5. The presence of node 9 in the optimal node sequence indicates that, in this case, the strategy of setting a waiting time at node 4 has been successful.

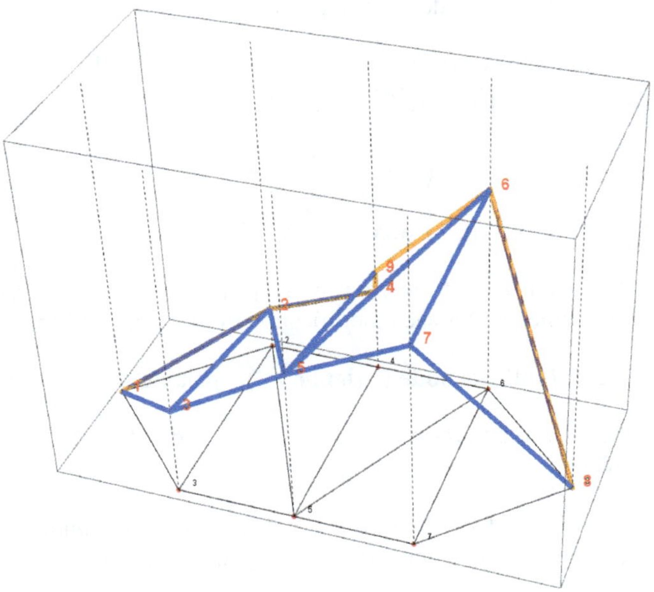

Figure 8: Optimal route connecting points 1 and 8 in 3D-space.

3 AN ADAPTED ALGORITHM

The algorithm presented below includes the necessary modifications to be able to determine the fastest paths between pairs of nodes within a network, with the weights of their arcs being subject to variations according to known or forecast schedules in advance.

Adapted Dijkstra's algorithm with time-dependent arcs

1. Read graph **G= (V, A),** adjacency matrix **Ady** and matrix **D** of displacement costs between pairs of adjacent points.

2. Build vector **f** that will store the minimum displacement costs to the origin from each node of **G** (initially, only the nodes adjacent to the origin will not have infinite value).

3. Build vector **p** that stores the successor nodes to each node of G in the optimal path (initially, only nodes adjacent to the origin can be determined).

4. Initialize set **S** of nodes explored by including only the origin point. **Initialize** set **LIST=G\S.**

5. While LIST is not empty:

5.1 Identify index **j***

$$j^* := Arg\left[\min\{f(j) : j \in LIST\}\right]$$

5.2 If all output arcs from node **j*** do not change in time **then**

5.2.1 Remove index j* from LIST

5.2.2 For each successor **k** of j* included in LIST:

If $f(k) > f(j^*) + d_{j^*k}(0)$ **then**

i. **Update** $f(k) := f(j^*) + d_{j^*k}(0)$

ii. **Update** $p(k) := j^*$

5.3 else For each output arc from **node j*** towards some not visited node that changes over time and for each modification of the weight of this arc:

5.3.1 Replicate node j* (let be j*'), updating set V.

5.3.2 Add an arc from j* to j*', weighted with the required wait time at node j*.

5.3.2 Repeat for the new node j*' the same out connections that node j* already had (with the corresponding new weights).

5.3.3 Add an arc from j* to j*', weighted with the required wait time at node j*.

5.3.4 Update set A according to 5.3.2 and 5.3.3.

5.3.5 Include node j*' in set LIST.

6. End

Note that the added programming block mainly corresponds to the section labelled 5.3, and it is activated when time-dependent output arcs are detected. The complexity of this algorithm, that maintains the same structure of the original Dijsktra's algorithm, depends precisely on the number of arcs whose weights are subject to time-dependent changes, as well as the number of changes to be incorporated along the time horizon for each arc.

4 CONCLUSIONS

In this article, the problem of determining the fastest paths with time-dependent arcs in transport networks has been analysed. There are numerous existing bibliographical contributions related to this subject, mainly due to the relevance that this issue acquires in logistics and travel planning for all types of users. Specifically, the methodology presented in Wen et al. [17], where two heuristic methods were proposed to solve the least cost path problem between a pair of nodes with a time-varying road network and a congestion charge. The tool developed by these authors was based on modifications of Dijkstra's algorithm, where a wait ban had been established on the nodes.

The fastest path algorithmic search technique with time-dependent arcs introduced in this contribution follows this methodological line of adaptation of Dijkstra's algorithm to this context, which guarantees a high level of efficiency for the calculation of solutions. On the other hand, the original graph must be conveniently extended, both in number of nodes and in new arcs (some of them unidirectional, which are those that indicate the time progress). It would be possible, however, to limit this growth of the solution container graph, having a reliable control of the time when the possible alternatives are analysed.

ACKNOWLEDGEMENT

This work was in part supported by the Ministerio de Investigación (Spain)/FEDER under grant PID2019-106205GB-I00. This support is gratefully acknowledged.

REFERENCES

[1] Dijkstra, E.W., A note on two problems in connection with graphs. *Numer. Math.*, **1**, pp. 269–271, 1959.

[2] Cormen, T.H., Leiserson, C.E. & Rivest, R.L., *Introduction to Algorithms*, MIT Press and McGraw-Hill: London, 1994.

[3] Toth, P. & Vigo, D., *Vehicle Routing: Problems, Methods, and Applications*, Vol. 18, SIAM, 2014.

[4] Pillac, V., Gendreau, M., Guéret, C. & Medaglia, A.L., A review of dynamic vehicle routing problems. *European Journal of Operational Research*, **225**, pp. 1–11, 2013.

[5] Preuss, T. & Syrbe, J.-H., An integrated traffic information system. *Proc. 6th Int. Conf. Appl. Computer Networking in Architecture, Construction, Design, Civil Eng., and Urban Planning (europIA '97)*, 1997.

[6] Delling, D. & Wagner, D., Time-dependent route planning. *Robust and Online Large-Scale Optimization*, Springer-Verlag, Berlin, pp. 207–230, 2009.

[7] Eglese, R., Maden, W. & Slater, A., A road timetable TM to aid vehicle routing and scheduling. *Comput. Oper. Res.*, **33**, pp. 3508–3519, 2006.

[8] Gendreau, M., Ghiani, G. & Guerriero, E., Time-dependent routing problems: A review. *Computers and Operations Research*, **64**, pp. 189–197, 2015.

[9] Perakis, A.N. & Papadakis, N.A., Minimal time vessel routing in a time-dependent environment. *Transp. Sci.*, **23**, pp. 266–276, 1989.

[10] Norstad, I., Fagerholt, K. & Laporte, G., Tramp ship routing and scheduling with speed optimization. *Transp. Res., Part C*, **19**, pp. 853–865, 2011.

[11] Helvig, C., Robins, G. & Zelikovsky, A., The moving-target traveling salesman problem. *J. Algorithms*, **49**, pp. 153–174, 2003.
[12] Jiang, Q., Sarker, R. & Abbass, H., Tracking moving targets and the non-stationary traveling salesman problem. *Complex Int.*, **11**, pp. 171–179, 2005.
[13] Sutner, K. & Maass, W., Motion planning among time dependent obstacles. *Acta Inf.*, **26**, pp. 93–122, 1988.
[14] Latombe, J.-C., *Robot Motion Planning*, Springer-Verlag: Berlin, 1990.
[15] Fujimura, K., Time-minimum routes in time-dependent networks. *IEEE Trans. Robot Autom.*, **11**, pp. 343–351, 1995.
[16] Dreyfus, S.E., An appraisal of some shortest-path algorithms. *Operations Research*, **17**, pp. 395–412, 1969.
[17] Wen, L., Catay, B. & Eglese, R., Finding a minimum cost path between a pair of nodes in a time-varying road network with a congestion charge. *Eur. J. Oper. Res.*, **236**, pp. 915–923, 2014.

SUSTAINABLE MOBILITY, TRANSIT-ORIENTED DEVELOPMENT, AND TRAVEL BEHAVIOR: A TRAVEL PATTERN ANALYSIS

AYA HASAN ALKHEREIBI
Department of Mechanical and Industrial Engineering, Qatar University, Qatar

ABSTRACT

This research proposes an inclusive travel pattern classification model, in support of sustainable mobility and transit oriented development to develop homogeneous activity groups for a sample of laborers in the State of Qatar. The investigated model aims to classify the activity data into a homogenous travel pattern. A pattern recognition model is applied to a revealed preference (RP) survey for the travel diary of 1,051 laborers. In the first phase of the analysis, raw data preprocessing algorithms and outliers data detection and filtering algorithms were applied and, therefore, an activity-based displacement matrix was developed for each household. The research methodology commenced in this research encompasses the integration of several machine learning (ML) techniques, mainly utilizing clustering and classification methods. A bagged clustering algorithm was used to recognize the clusters' number, and then the implemented CMeans algorithm and Pamk algorithm were used to validate the results. Meanwhile, the interdependencies between the resulting clusters and the socio-demographic characteristics of the household were examined using cross-analysis. The results of the study found that there was a notable diversity between clusters in terms of trip purpose, modal split, choice of destination, and occupation. Clustering techniques on all three attributes produced similar results, but clustering based on occupation yielded clusters that differed significantly from those based on other attributes. Applying such pattern recognition models to large and complex activity datasets could help transportation planners better understand the travel needs of segments of the population and formulate more informed strategies that compromise the best practices of sustainable mobility and transit-oriented development.

Keywords: travel behavior, transportation planning, activity-based model, machine learning, support vector machine, blue collar travel diary.

1 INTRODUCTION

1.1 Overview

In contemporary city planning, the use of public transportation is often considered one of the most significant strategies to reduce traffic congestion and pollution [1]. However, the snag of this approach is that people are less likely to walk around their neighborhoods and socialize. To counter this, transit-oriented development (TOD) has become one of the most popular urban planning approaches in recent decades [2]. This type of urban development aims to promote the use of mass transit by providing an environment where residents live within walking distance of a major through station and other amenities. It's expected that this will increase ridership and the use of public transport while allowing access to more jobs, education, and cultural facilities [3].

TOD is also aimed at providing a pedestrian environment and better accessibility, which promotes pedestrian activities [2]. Mentioning residents' activities, implement machine learning classifiers to recognize activity-based travel behavior patterns for a sample of laborers who represents a significant portion of residents of the Qatari society. This research problem is tackled from a dissimilar perception by developing a classification activity-based model that groups the homogeneous daily activity data into clusters. Each cluster has its essential features such as the type of activity, sequences, start time, activity duration, etc.

Moreover, the conventional trip-based travel demand modeling approach has been criticized for its conceptual deficiencies and lack of sensitivity to several important factors, including congestion, the distribution of activities over space, and the effects of economic, gender, social, transportation network, and locational divisions on spatial and temporal aspects of individual mobility [4]. The activity-based approach views travel as a derived demand; derived from the need to pursue activities distributed over space [5]. The approach focuses on sequences or patterns of activity behavior, with the whole day or longer periods of time as the unit of analysis. It emphasizes the effects of (and constraints imposed by) economic, gender, social, transportation network, and locational divisions on spatial and temporal aspects of individual movement.

The activity-based holistic perspective of travel can reliably evaluate urban travel demand management policies since it explicitly models activity patterns and considers these patterns to be the fundamental influence on travel. The objective of this ongoing research is to contribute toward advancing and operationalizing the activity-based approach by (a) Examining and understanding activity-based travel behavior, (b) developing a comprehensive representation of attributes characterizing laborers' daily activity-travel pattern, and (c) proposing an analysis framework for modeling the activity-travel pattern of individuals based on careful descriptive analysis of data, d) based on the defined travel pattern the research proposes TOD strategies that enhance smart mobility practices in the state of Qatar.

1.2 Literature review

Machine learning methods have been a boon to the transportation industry. They've aided planners in identifying patterns in travel data and using them to predict future travel behavior. Recently, Allahviranloo & Aissaoui [6] applied k-means clustering to identify distinctive travel clusters. They applied the AdaBoost algorithm to predict the start time and duration of travel-based activity based on individuals' sociodemographic features and individuals' daily schedules [7]. Researchers have used interpersonal variability to generate intrapersonal variability, but the new technique introduces a data-driven activity scheduling approach, which swaps several expert-designed mechanisms and their complex engineering interactions for a handful of machine learning techniques. Drchal et al. [8] noted that the new technique is more efficient than previous methods, as it does not require extensive engineering efforts or extensive data collection. They also noted that it is more accurate than previous methods, as it has a higher ability to predict human behavior. In order to execute this technique, the authors used an optimization algorithm based on a genetic algorithm (GA) and a particle swarm optimization (PSO). The resulting model is capable of generating an accurate schedule in real-time while being able to adapt dynamically to changes in traffic conditions or other external circumstances.

Activity-based travel demand models have made significant advances in the past decade, but there are still some significant opportunities for improvement. Despite advances in generating temporal information for the design of activity-based transport demand models, there is significant scope and opportunity to improve the performance of these models in terms of estimation accuracy, estimation efficiency, computations, and their impact on case studies.

The activity-based modeling is a technique that can be used to identify patterns in the activities of people on a daily basis. This model has been utilized by several scholars such as

Hafezi et al. [9] and Awais Shafique and Hato [10]. The key idea behind activity-based modeling is that it can be used to model the temporal dimensions allied with the traveler's daily activity patterns, which have an important impact on the travel experience. The random forest algorithm was proposed by Hafezi et al. [9]. It uses a random forest model with CART techniques for measuring features' importance. However, this technique has also been used in other areas such as traffic accident detection and means of transport [10]. In their study, using activity and temporal models, Hafezi et al. [9] improved current activity-based techniques by using daily data and pattern recognition models to identify groups of people who have ongoing demands for their activities outside of their place of residence. And more broadly, similar travel patterns along with the type of activities and start/end times along with the duration.

The SALT planner framework uses a set of advanced rule-based econometric models to predict daily trip demand specifications and route generation. Identifying homogeneous populations can significantly improve the estimation accuracy of activity-based transport demand models. Moreover, the SALT planner framework is designed to obtain a dynamic model for traffic demand in an urban area by using both time-series data and spatial information. It uses an advanced combination of traffic forecasting techniques through an econometric model to predict daily trip demand specifications and route generation.

2 METHODS AND MATERIALS

Table 1 shows the general characteristics, in terms of age, income, education level, and occupation, of the investigated sample.

Table 1: Sample general characteristics.

Sample characteristics
Age
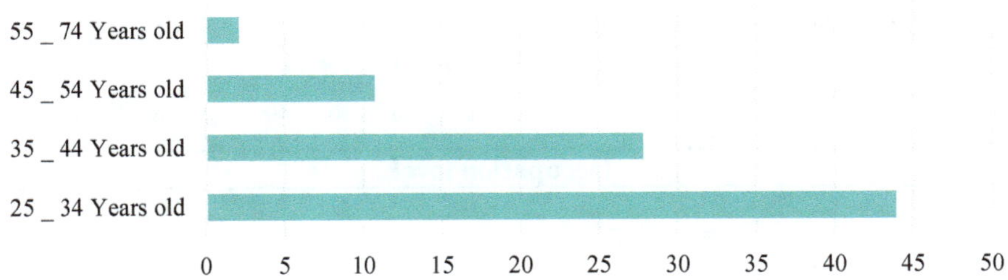
Income
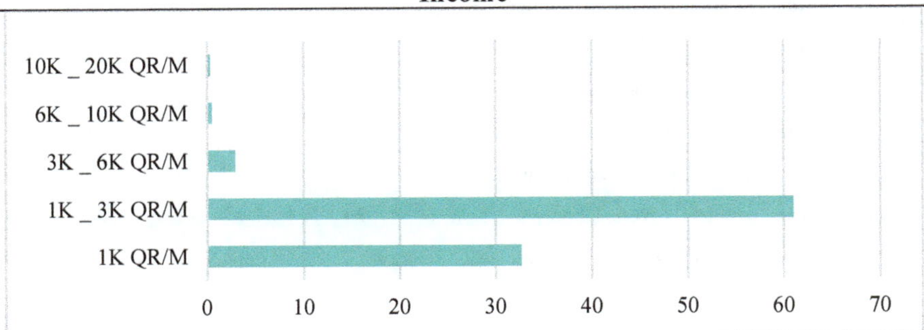

www.witpress.com, ISSN 1743-3509 (on-line)

Table 1: Continued.

Sample characteristics
Education

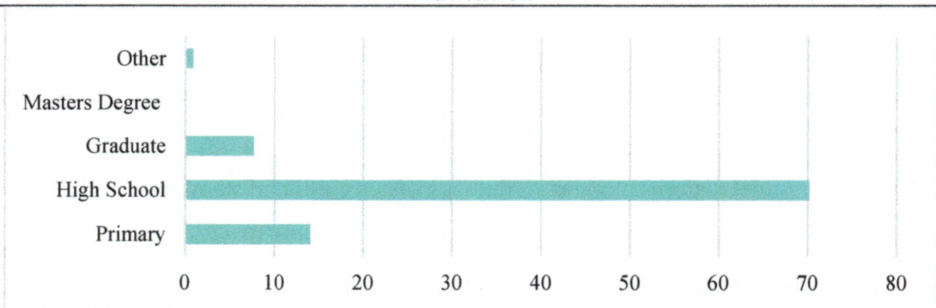

Occupation sector

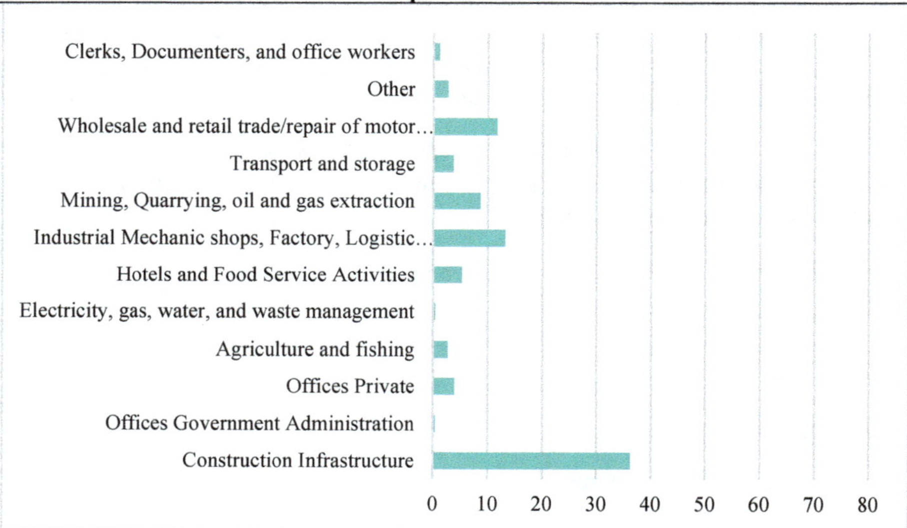

Occupation level

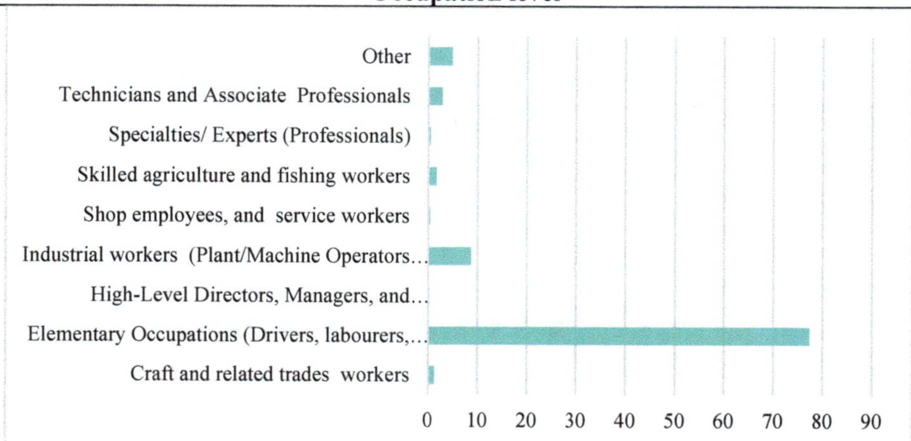

Age: More than 93% of the respondents were from the category of economically active ages ranging from 19 to 54. Similarly, the overwhelming majority of the respondents (93%) earn QAR3,000 ($USD 820) or less per month. The education level of respondents is mostly secondary and high school educators with a total percentage of 70.2.

Income: Almost all (61%) respondents indicated that they earn no more than QAR3,000 per month while 3% earn between QAR3,001–6,000 per month; 1% earn between QAR6,001–10,000 per month, and 0.3% earn more than QAR10,001 per month.

Activity-based travel patterns (ABTP) are travel recognition pattern recognition that is based on activities that may change over time. The recognition and comprehension of activity-based travel patterns are important for developing efficient transportation systems [6], [11]–[13]. One method used by Leisch [14] was the bagged clustering (B-Cluster) algorithm for initialization of the number of clusters. Subsequently, C-Means and Pamk algorithms were applied to optimize the performance of the B-Cluster algorithm as recommended by Hung and Yang [15]. In this study, we used the CROSSTABS test to deal with interdependencies and applied the CART classifier algorithm for the alignment of a person's socio-demographic characteristics with predefined clusters. The theoretical model proposed herein has the capability of learning and predicting temporal information of patterns of activity for individuals concerning the characteristics of heterogeneity. Besides, applying this framework to activity-based models reveals the effectiveness of machine learning techniques in distinguishing temporal attributes of travelers' daily activity patterns, but also helps better understand the relationship between travel planning and patterns of activities with respect to heterogeneity characteristics.

A flowchart of the research methodology followed in this study is displayed in Fig. 1.

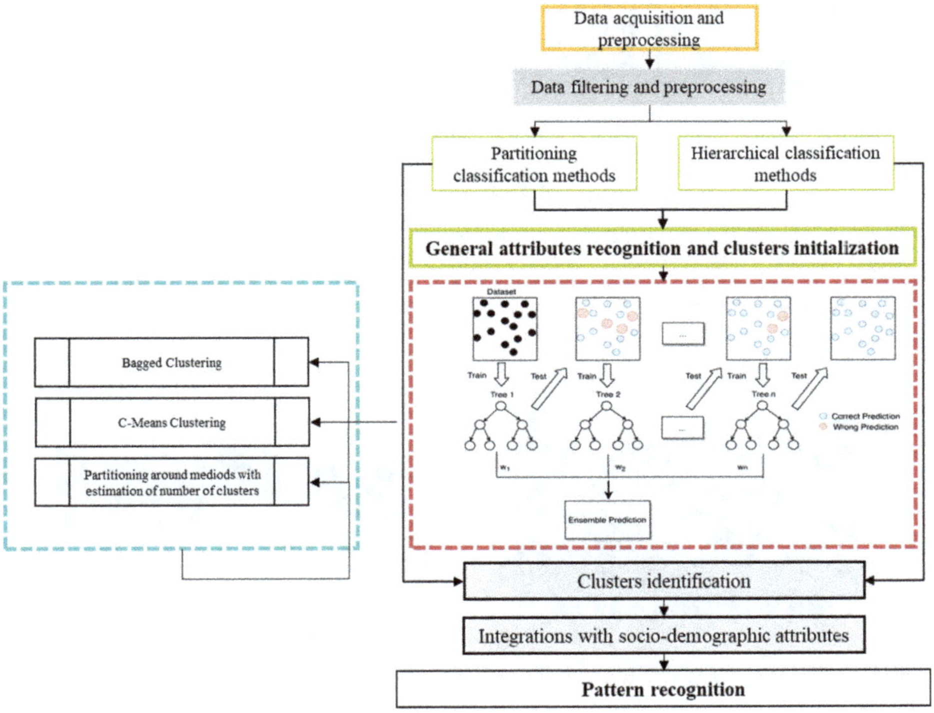

Figure 1: Research design flowchart.

3 RESULTS AND KEY FINDINGS

Different structures are observed by applying different grouping techniques. Table 2 describes the results of these clustering techniques. Figs 2–4 show the plots for the three clustering techniques. As described, it appears that Cmeans and Pamk produce an even distribution compared to the bag clustering technique. The latter yields insignificant records for clusters 5, 6, 7, 9, and 10, as well as null Logs for Cluster 8.

Table 2: Number of clusters per method.

Method	Cluster										
	1	2	3	4	5	6	7	8	9	10	Total
Bclust	2,150	2,405	28	356	32	3	4	0	8	4	4,990
Cmeans	358	583	43	658	353	612	404	1,369	572	38	4,990
Pumk	604	1692	693	443	591	529	387	38	8	5	4,990

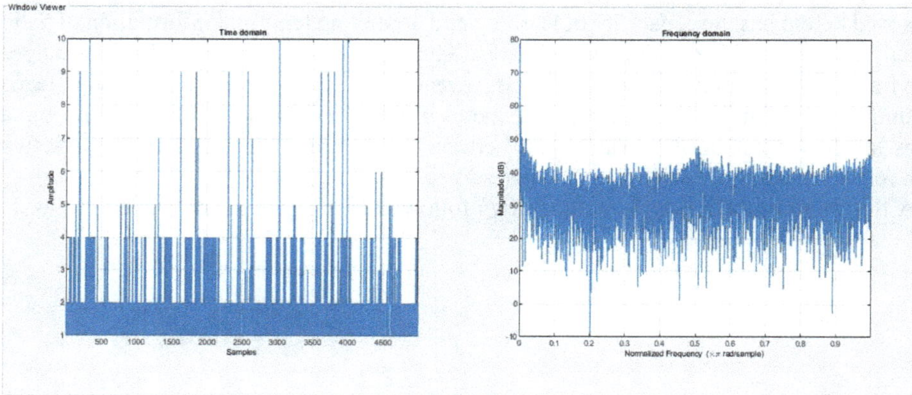

Figure 2: Bagged clusters.

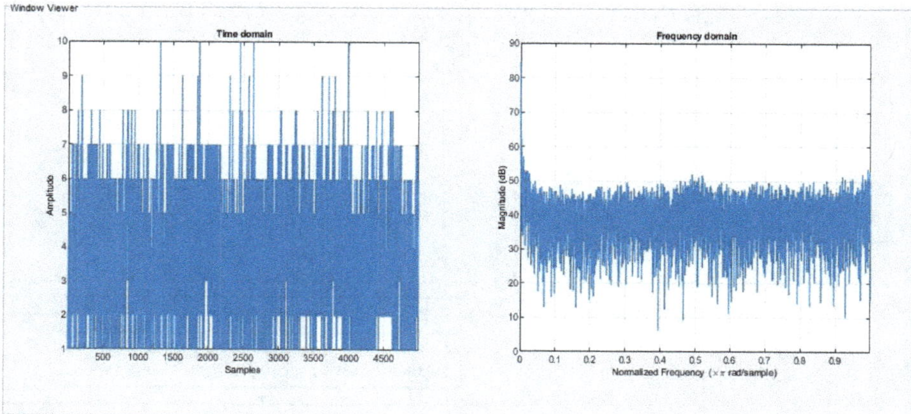

Figure 3: C-means clusters.

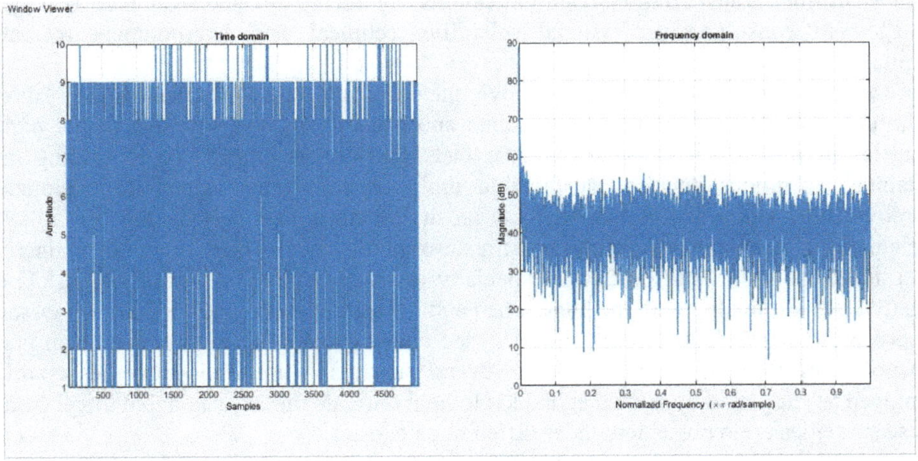

Figure 4: Pamk clusters.

The C-means clustering technique has been used to cluster the activities of laborers based on their work, job duties, and personal activities. According to the results, it is observed that the three main activities are work and job duties, personal activities, and preparing meals for a group of co-workers. These have shares of 21%, 43%, and 31% respectively. Engagement in other activities such as study is negligible. Then the Bagged clustering technique was used to create a set of clusters from observations with similar attributes; this is done by combining several methods of clustering into one. This combines the advantages of each individual method and reduces the disadvantages of each individual method. Utilizing the bagged clustering gave the same characteristics of cluster 1 as the C-means technique.

In the context of traveling to work, it is possible to make a distinction between three main modes: returning home, commuting to the workplace and traveling to other places. By the way, Al Rayyan, Al Sheehaniya and Doha are respectively three main municipalities where trips were made.

Al Rayyan is the municipality with the highest share of trips in all categories (i.e. working trips, leisure and shopping trips). On the other hand, Al Sheehaniya is second in terms of transportation modes (i.e. private shuttle and car/van/truck) whereas Doha is ranked third for both categories (i.e. private shuttle and car/van/truck).

In terms of income, the majority of earners make between QR 1,001 and QR 3,000 per month (around 63%). Around 33% earn less than QR 1,000 monthly; 2.5% earn between QR 3,001 and QR 6,000 per month; only 1% earns more than QR 6,000 monthly every month.

The majority of the workers in this cluster were Indian (43%) followed by Nepalese, Bangladeshi and Filipino nationals respectively. Around 40% of the workers were between 25 and 34 years old. The other demographics include a high percentage of male workers at 98.9% and full-time employed adults at 99.9%. Nearly 84% of all occupations are classified as elementary occupations such as drivers and laborers while 9% are factory or industrial workers such as plant/machine operators and assemblers. Approximately 3% are technicians and associate professionals.

The main domain for these occupations is construction infrastructure (30%), industrial mechanics shops (20%), logistics manufacturing (10%), mining quarrying oil gas extraction (10%) motor mechanics (10%). Other domains include agriculture fishing forestry social services utilities transportation communications commerce finance insurance real estate management consulting professional scientific technical arts entertainment recreation hospitality.

According to the results presented above, the dominant transport mode used by laborers is the private shuttle. As stated by Shaaban and Maher [16], Qatar is one of the highest countries in private car ownership, these facts created the emergence to finding more sustainable solutions such as sustainable mobility and transit oriented development. Moreover, this study shows that there is an urgent need for studying the feasibility of implementing sustainable mobility and transit-oriented development in order to increase accessibility to public transportation, especially in the presence of the poor use of PT; less than (3%). Additionally, with the huge urban and infrastructure evolution Qatar witnessed in the preparation for FIFA World Cup 2022, the country had a huge development in public transportation as well. In this context, several sustainable mobility strategies could be employed as suggested by Hafezi et al. [13] to help mitigate the unsustainable travel pattern. These strategies are represented under different categories:

- Transportation: by increasing the use of public transportation, carpooling, and ride sharing, and reducing the distance between home and work by the well-planned allocation of laborers camps.
- Infrastructure: includes developing more walkable communities with bike trails and other amenities that encourage walking and biking. It also includes converting unused spaces into bicycle parking areas and creating better bicycle paths, and providing a shaded path to enhance walkability.
- Cultural awareness: by educating the public and especially laborers about the importance of sustainable mobility practices through campaigns.

4 CONCLUSION

This research presented the applicability of machine learning classifier models to an activity-based model for a sample of laborers in Qatar, to understand laborers travel patter then help decision-makers formulate strategies for the best practices of sustainable mobility and transit-oriented development. The proposed model applied clustering techniques to consolidate household activity data into a homogeneous activity model.

This study was undertaken to extract socio-economic and household travel information from a revealed preference (RP) survey collected by the Ministry of Transport and Communications (MOTC) of Qatar, which was used to uncover patterns of residents' activity, particularly in the context of trip purpose, mode choice and destination. Three clustering techniques (bagged clustering, C-means clustering, and Pamk clustering) were applied separately to uncover patterns of residents' activity in terms of trip purpose, mode choice and destination. The findings from this research study prove that the above three clustering methods are credible and could provide a better understanding of the residents' shift pattern. However, the CMeans and Pamk clustering technique show better efficiency and representativeness than the bagged clustering technique. The study clustered a large number of activities collected from travel diaries into meaningful clusters using machine learning techniques to best support the further development of predictive models for diverse elements of the transportation system Travel patterns have been identified which help to develop effective decision support systems in transit-oriented development planning and evaluation of different strategies. The results of the study showed that cluster analysis techniques are mathematically efficient and can classify residents into groups and, therefore, analyze travel behavior. By applying clustering techniques, detailed traveler characteristics such as income, gender, age, nationality and occupation are considered for activity-based models in addition to mode/destination choice. The identified travel patterns are valuable know-how and absorption in the development of effective decision-making systems in the sustainable mobility and transit-oriented development planning and evaluation of different strategies.

REFERENCES

[1] Chan, W.C., Ibrahim, W.H.W., Lo, M.C., Suaidi, M.K. & Ha, S.T., Sustainability of public transportation: An examination of user behavior to real-time GPS tracking application. *Sustainability (Switzerland)*, 12(22), pp. 1–20, 2020. DOI: 10.3390/su12229541.

[2] Thomas, R., Pojani, D., Lenferink, S., Bertolini, L., Stead, D. & van der Krabben, E., Is transit-oriented development (TOD) an internationally transferable policy concept? *Reg. Stud.*, **52**(9), pp. 1201–1213, 2018. DOI: 10.1080/00343404.2018.1428740.

[3] Nasri, A. & Zhang, L., The analysis of transit-oriented development (TOD) in Washington, D.C. and Baltimore metropolitan areas. *Transp. Policy (Oxf.)*, **32**, pp. 172–179, 2014. DOI: 10.1016/j.tranpol.2013.12.009.

[4] Moeckel, R., Kuehnel, N., Llorca, C., Moreno, A.T. & Rayaprolu, H., Agent-based simulation to improve policy sensitivity of trip-based models. *J. Adv. Transp.*, **2020**, 2020. DOI: 10.1155/2020/1902162.

[5] Joubert, J.W. & de Waal, A., Activity-based travel demand generation using Bayesian networks. *Transp. Res. Part C Emerg. Technol.*, **120**, 102804, 2020. DOI: 10.1016/j.trc.2020.102804.

[6] Allahviranloo, M. & Aissaoui, L., A comparison of time-use behavior in metropolitan areas using pattern recognition techniques. *Transp. Res. Part A Policy Pract.*, **129**, pp. 271–287, 2019. DOI: 10.1016/j.tra.2019.08.007.

[7] Li, Y., Wang, C. & Gong, J., A multi-model probability SOC fusion estimation approach using an improved adaptive unscented Kalman filter technique. *Energy*, **141**, pp. 1402–1415, 2017. DOI: 10.1016/j.energy.2017.11.079.

[8] Drchal, J., Čertický, M. & Jakob, M., Data-driven activity scheduler for agent-based mobility models. *Transp. Res. Part C Emerg. Technol.*, **98**, pp. 370–390, 2019. DOI: 10.1016/j.trc.2018.12.002.

[9] Hafezi, M.H., Daisy, N.S., Millward, H. & Liu, L., Ensemble learning activity scheduler for activity based travel demand models. *Transp. Res. Part C Emerg. Technol.*, **123**, 102972, 2021. DOI: 10.1016/j.trc.2021.102972.

[10] Awais Shafique, M. & Hato, E., Use of acceleration data for transportation mode prediction. *Transportation*, **42**, pp. 163–188, 2015. DOI: 10.1007/s11116-014-9541-6.

[11] Cui, Y., He, Q. & A. Khani, A., Travel behavior classification: An approach with social network and deep learning. *Transp. Res. Rec.*, **2672**(47), pp. 68–80, 2018. DOI: 10.1177/0361198118772723.

[12] Chen, S., Prakash, A.A., de Azevedo, C.L. & Ben-Akiva, M., Formulation and solution approach for calibrating activity-based travel demand model-system via microsimulation. *Transp. Res. Part C Emerg. Technol.*, **119**, 102650, 2020. DOI: 10.1016/j.trc.2020.102650.

[13] Hafezi, M.H., Daisy, N.S. & Millward, H. & Liu, L., Ensemble learning activity scheduler for activity based travel demand models. *Transp. Res. Part C Emerg. Technol.*, **123**, 102972, 2021. DOI: 10.1016/j.trc.2021.102972.

[14] Leisch, F., Bagged clustering. *Adaptive Information Systems and Modelling in Economics and Management Science*, **51**(51), p. 11, 1999. DOI: 10.1.1.33.642.

[15] Hung, M. & Yang, D., *An Efficient Fuzzy C-Means Clustering Algorithm Ming-Chuan,* IEEE, 2001.
[16] Shaaban, K. & Maher, A., Using the theory of planned behavior to predict the use of an upcoming public transportation service in Qatar. *Case Stud. Transp. Policy*, **8**(2), pp. 484–491, 2020. DOI: 10.1016/j.cstp.2019.11.001.

SECTION 2
MOBILITY IN THE URBAN ENVIRONMENT

CONCEPT OF ACCESSIBILITY IN SUSTAINABLE TRANSPORT: CRITERIA AND PERSPECTIVES

HODA POURRAMAZANI & JOSE L. MIRALLES-GARCIA
Department of Urbanism, Polytechnic University of Valencia, Spain

ABSTRACT

In today's world, accessibility is considered a key and complementary criterion while increasing the stability of a transportation system, in addition to the importance of proper density and spatial distribution of people and activities and land use development, which is known as urban policy and services and it is also considered a precondition for economic development. Although accessibility has been extensively explored in recent decades, transportation planners still face many challenges. Therefore, it is important to know the factors and criteria affecting access in each time period. These criteria can include mobility, quality and cost-effectiveness of transportation, the connection of transportation systems, mobility alternatives, land use patterns and socio-economic characteristics that can affect the concept of access differently; on the other hand, a more comprehensive study of accessibility can increase the range of potential solutions to transportation problems. The present study provides a general but comprehensive overview of the various dimensions and indicators of access in the form of a literature review. The results show that there is a lot of interaction with the goals of integration and sustainability, which can be achieved under the key elements. The first element is the selection of techniques that can be presented in the light of evaluation along with specific and clear objectives that can be achieved in practice and reflect their ease. The second element is the use of techniques to provide sufficient indicators according to regional patterns, taking into account all objectives, of course, as general performance indicators that are not limited to social justice. Perhaps one of the effective options is the use of visualization tools. It means using maps that provide solutions. It provides a link between transportation gaps and benefits where access-based criteria can potentially influence transportation decisions. Also, the findings show that to strengthen the accessibility goal, accessibility indicators should be clearly considered, and a distinction should be made between this concept and mobility in transportation planning.

Keywords: accessibility, criteria, transportation, land use, socio-economic dimensional.

1 INTRODUCTION

Accessibility is a concept of continuous relevance in transportation research and has been raised as an essential goal in planning in recent years. This concept is increasingly seen as an alternative to the planning and directional mobility paradigm [1] because it allows for the complex interactions between land use, transport and socio-economic conditions to provide a wholesome social view of transport [2].

It is also stated that better access reflects the socio-economic benefits of the network, that is, in terms of land values and quality of life. Because it seeks to establish fair access for all population groups to additional services such as health care [3]. Access to transportation is an essential driver of urban growth and the key to the sustainable development of cities. The growing interest in this development has emphasized the importance of accessibility as a key indicator for evaluating urban transport investment and urban form; that is, it can link land use and transport together and thus provide a basis for comprehensive and integrated policies to include coordinated actions on a daily basis simultaneously [4]. Accessibility to businesses, services, and other destinations of interest has long been recognized as key to sustainable transportation development and essential spatial planning strategies [5]. Increasing emphasis is also placed on this key indicator of economic development, as the movement of people, goods, and environmental goals such as reducing emissions of

WIT Transactions on The Built Environment, Vol 212, © 2022 WIT Press
www.witpress.com, ISSN 1743-3509 (on-line)
doi:10.2495/UMT220051

greenhouse gases and pollutants from different modes of transportation and how they are used to support economic performance, it is considered essential and effective. Hence it makes effective performance in the economy possible [6].

On the other hand, based on accessibility and its interaction with sustainability, Hansen [7] shows a strong correlation between the accessibility of an area within the city and its ability to attract urban development or investment and is stated that with regard to social equality, economic, development and environmental effects of accessibility as a key element in urban sustainability should be considered.

2 METHODOLOGY OF THE LITERATURE REVIEW APPROACH

A systematic literature review and articles were used to complete this article that was selected based on their importance, relevance to the purpose of the article, and publications of recent years have been used, and then a brief overview of the obtained results was made. This selection of bibliographic sources is based on authoritative publications such as Marwal and Silva [8], Kelobonye et al. [9], Siddiq and Taylor [10] and Bivina et al. [11]. It can be said that one of the most important articles in the field of accessibility belongs to Marwal and Silva [8], who reviewed the accessibility literature in the last 5 years. However, according to the purpose of this article, which is not a comprehensive review of all articles related to access, and considering that the evolution of cities is always accompanied by many challenges and in recent years, it is moving more and more toward achieving sustainability. The concept of accessibility is not an exception to this rule and is changing with a new definition, so based on sustainability goals, we seek to know the new era of accessibility and its indicators. In this regard, from the key concepts (access, access criteria, sustainable mobility) based on the topic, a research was conducted in Google Scholar, Gate Researcher, Scopus or Academia.edu to obtain the necessary information. Therefore, the search was limited by the article's aim to collect approximately 107 publications (e.g., articles, books, dissertations). The selected publications mainly focus on documents published since 2007 (Fig. 1).

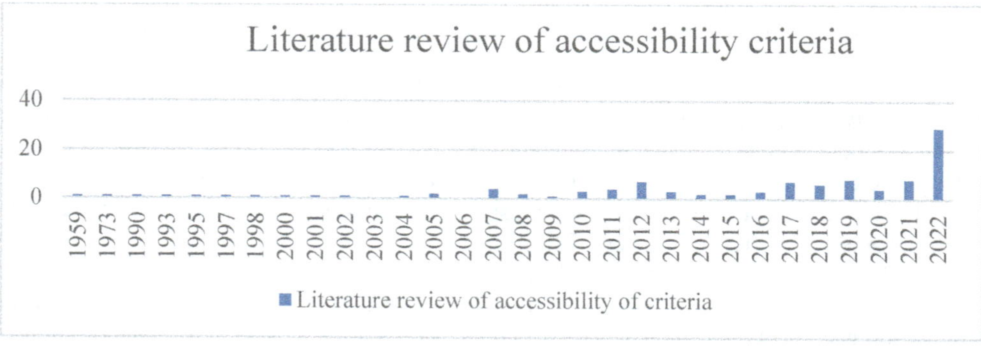

Figure 1: Year of publication of the 107 documents selected in the literature review of accessibility criteria.

3 THE CONCEPT OF ACCESSIBILITY AND ITS TYPES

There are several definitions of accessibility in the literature. However, a general definition by Wachs and Komagai [12] states that accessibility is the ease or difficulty with which opportunities (for example, employment) or services can be obtained from a place. Geurs and van Wee [13] define accessibility as the ability to reach activities or destinations by an individual or a combined mode of transportation. Accessibility is the effort necessary to overcome spatial and spatial separation (e.g., Paez et al. [48], Merlin and Hu [27], Litman [15], Halden [22]), and it usually reflects the utility (e.g., home to work) associated with travelling to places [14].

Accessibility is defined as the potential to achieve distributed opportunities (employment, recreation, social interaction, etc.), and it can be considered one of the primary spatial outputs and the joint result of the transportation network and the geographical distribution of activities (e.g., Geurs and van Wee [13], Papa and Coppola [25], Lucas et al. [46], Pajares et al. [36]), In other words, accessibility can be seen as the ease of obtaining services and activities [15]. According to this definition, accessibility depends to a large extent on the spatial distribution of destinations, the land use component and the ability to move from one place to another, and the transportation component [13]. The concept of access in a suitable framework for achieving sustainable land use and transportation system and environmental and climate concerns has been of special importance (e.g., Kinigadner and Büttner [44], Banister [42], Levine et al. [41], Siddiq and Taylor [10]). This concept provides a suitable framework for integrated land use and transport planning and is a key factor in achieving sustainable mobility [16], [17]. Based on one of the long-standing concerns of researchers, they find it is useful for informing basic urban planning and policies (e.g., Proffitt et al. [20], Halden [22], Delbosc and Currie [47]).

3.1 Types of accessibility

According to the literature, accessibility can be divided into three categories: accessibility to public transportation, auto accessibility, and transportation. As already mentioned, accessibility has been raised as a significant goal in recent years, but one of the most important challenges is the lack of sufficient knowledge about public transport and automated access, even when who decide where to live and value it differently (e.g., Foth et al. [3], Ariza-Álvarez et al. [49], Levine et al. [41], Cervero et al. [31], Nasri and Zhang [29]).

Despite this, it has been stated that all three types of access are essential determinants in choosing a residential location, and transit access has a statistically significant and positive effect. In the case of automatic access, the results show that commuting time with automatic mode has the most significant impact among these three variables and is independent, but access to non-work destinations seems to be of little importance (e.g., Næss et al. [32]). Research also states that due to restrictions such as housing supply, households are often forced to live in neighbourhoods with a lower level of accessibility than they prefer.

This point shows the undeniable influence of access in the culture of people's lives. Therefore, transportation planners should pay more attention to this concept and pay attention to existing inconsistencies. From an equity point of view, comparisons between types of access for different modes of transport are needed [18] because, on the one hand, the car tends to provide significant levels of access in most urban areas. On the other hand, according to the concept of sustainable urban development, access to cars due to the globalization of car ownership and use barriers (costs, legal requirements and physical ability) and in the environmental dimension (greenhouse gas emissions and energy consumption). Has undergone many changes, especially in recent years with the emergence of alternative methods. Therefore, it is necessary to compare types of access at the regional level because it provides insight into the level of access of people or households without cars at a level comparable to people who own cars in the same area.

4 MOBILITY AND ACCESSIBILITY

A growing number of transportation planning researchers and practitioners are discussing the shift from mobility-based planning to accessibility-focused planning, as accessibility is an alternative approach to mobility-based planning. Accessibility is a compelling concept that is more difficult to operationalize than mobility, which many accessibility criteria have been developed to explain, thereby making accessibility-based planning more efficient and taking into account active public transport and inclusion. These criteria use policies that reduce the distance between activities [19].

Accessibility captures the complex interactions between different perspectives, while mobility simply reflects the ease of movement and ease of reaching desired destinations, which is actually the reason people want to travel [20]. Accessibility in the literature is one of the most comprehensive measures to evaluate a region's complex functions of land uses and transportation systems. In this regard, it has been mentioned as a factor in reducing the use of cars, which increases social equality and supports economic development. In a study to quantitatively assess America's transportation, Proffitt et al. [20] found that accessibility is increasingly included in plans, especially in large metropolitan areas, but is still marginally addressed compared to mobility.

Similarly, in an evaluation of four plans in California, Handy [21] found that programs based on mobility were created. However, these plans also addressed some concerns about access. In the UK, accessibility was also widely discussed and used, mainly due to the creation of accessibility programs that were designed and implemented at the behest of the government [22]. But still, with all interpretations, the practical implementation of limited access remains.

Another article evaluated 32 urban transportation plans from North America, Europe, Asia and Australia. According to the goals and performance indicators of access, the results showed that there is a tendency towards greater integration of access. However, still, few plans showed access-based indicators that can guide the decision-making process. Therefore, it has been stated that by strengthening the approach based on equal access to transportation, programs should clearly define the goals of access and differentiate between access and mobility. Hence, indicators need to guide the decision-making process [23].

5 ACCESSIBILITY CRITERIA

In the literature, many researchers have studied access indicators. For example, many of the indicators identified by Handy and Niemeier [24] were still valid at the time of the Geurs and van Wee review [13], including the need to improve land use and transportation assessment. Accessibility index in a wide base of theoretical, operational, interpretability and communication capabilities open the way to be used for evaluation purposes. Some of these criteria are normative and are justified according to how much people should travel or how much travel is reasonable for people. Some are based on positive accessibility that can be understood in terms of the distance that people actually travel, or in other words, normative accessibility. It is defined in terms of an expectation from an analyst or policy maker, but positive access is defined as the actual experiences of people who pass through the space to engage in outdoor activities.

What has been shown in the literature, accessibility with four prominent indicators, can be seen and investigated in all studies. Geurs and van Wee [13] illustrate the indicators well and consider and distinguish four accessibility criteria: transportation index, land use index, time-based index and person-based component. Some of these indicators include several other subcategories to be complete, so it should be noted that the research and application of these indicators are fundamental. On the other hand, it requires sufficient knowledge and awareness in this field.

5.1 Criterion based on location and land use

The first discussed index is the location-based index which includes land use and housing. This component includes the distribution of different types of uses in space, which is defined in terms of quantity (residential density, employment) and quality (level of employment, value of housing or the importance of services such as large hospitals and educational and cultural institutions). Measures based on this index provide insight into locations (for example, the number of jobs within 30 minutes of the place of origin).

This measure is mainly used by policymakers because it provides comprehensive measures of land use and transportation at the regional level and by location. As a central indicator of location advantage, this index is the primary measure of land use and transportation performance; on the other hand, accessibility based on this criterion has become a fundamental issue in fields such as urban planning, geography, and transportation economics.

It is also stated that the location-based index can facilitate land use and transportation, and policymaking in integration with each other. This index is able to analyze land use and transportation. By using this index, the access levels in different spatial units are specified and are comparable, so the lack of access, potential land use and transportation solutions can be deduced [16]. Based on this, the number of opportunities that can be created from a specific location is usually calculated based on the cost of travel to the destination using a particular mode of transportation.

The relationship between this criterion and accessibility has been strongly positive and significant, and the relationship of other accessibility variables of this index has been estimated as marginally significant. The results also show that in places where access to public and automatic transportation has a substantial effect on choosing a residential place, the automatic commuting time has the most significant effect on the choice, and the calculation of travel times that It can also include the mode index, often based on existing data. Therefore, this index can be generalized to the following set, which is very important and practical in examining actions based on it:

5.2 Gravity-based criterion

An attraction-based measure where all travel opportunities and costs generally decrease based on travel time or distance. This measure better reflects the journey, which is very suitable for travellers' understanding of time. From a scientific point of view, this criterion has a strong theoretical basis that is well accepted in the field of transportation. In addition, the spatial modelling framework considers the distribution of demand and the level of access to opportunities (competition effects). In principle, it can consider the changes between people. In other words, according to the characteristics of the people who are used, access is determined differently [25].

Research indicates that this measure is theoretically more correct (because it is not limited to a time or distance limit) and therefore has superiority over the measures of cumulative opportunities [10], [13]. It has been stated that this criterion identifies and smooths difficult destinations by using travel behaviour to make access easier. But by weighting opportunities based on distance or time, it makes the task easy and comprehensible. It can be understood as equivalent to the number of jobs compared to the cumulative opportunities that represent destinations at the one-time (or distance) predefined interpretation [26].

5.3 Cumulative opportunities

Cumulative opportunities only count opportunities that are within a specific range, and within the threshold of attraction-based travel costs, i.e., it estimates achievable opportunities at a predetermined or defined threshold. This index is essential from the perspective of policy and practice, even evaluating the positive impact of any type of access for choosing a residential location, for at least two reasons: it can clarify the debates related to budget allocation. Second, the investigation shows the different extent that people value access to different travel methods. Cumulative opportunity is the broadest measure in operational planning, but at the same time, it is among the smallest due to the lack of consideration of the effects of competition for opportunities. This criterion is of interest due to its ease of interpretation, reproducibility and direct comparison between cities [9], [27].

But one of its limitations, which is widely accepted in politics, is its failure to combine the effect of competition for existing opportunities. Because this criterion assumes that people use the closest opportunities and more to choose a residential place, further, it is assumed that there is no limit to the capacity of these opportunities. Therefore, the concept of more opportunities and more choices without considering competition or potential demand for opportunities can be misleading [28].

5.4 Criterion based on travel mode

When studying the location-based index, we come to another measure called travel mode. Since private cars, public transport and active are the most used modes of travel, the types of access that are usually investigated originate from these modes, especially in the field of transport choice. It is also important to note that studies that use location-based access indicators to assess access differentially by car and transit provide relatively simple estimates of mode-based travel time to compare levels, and it should be noted that these indicators are at the spatial scale of analysis. They are usually much larger than a typical study focusing on neighbourhoods or transit areas, and the calculation of mode-based travel times is often based on available data from transit demand models and schedules.

The paper reported that the share of the location-based index was higher for residents who lived near significant rail transit stations in 35 metropolitan areas across the United States. They examined the built environment and transportation accessibility at the neighbourhood and regional levels to determine the relationships between accessibility and commuting mode share. They showed that at the regional level, the level of access to jobs and transportation and the overall shape of the city affects how much people use each mode of transportation [29].

This is one of the articles that analyze the contribution of the travel mode at the national scale with the location-based index and the built environment. Of course, several studies have been conducted in the past that show that the development of mixed land use with high density and the choice of travel mode affects the way people commute to/from the workplace [30]–[32].

Accordingly, Kaufman et al. [33] found in his studies that car use was mainly explained by the local and sub-regional built environment (e.g., walkability, access to work by bus) rather than rail access and Efforts to activate dense and mixed areas in places beyond the areas adjacent to the railway station should be considered.

5.5 Proximity-based criterion

In the meantime, some researchers point to another new criterion known as the proximity index. This index is based on proximity to key destinations such as city cores and proximity to transportation infrastructure such as transit stations. This index is strongly associated with walking in many places, while it is generally assumed that households generally prefer lower density living. Several residential location selection models have reported a positive relationship between population density and this index and residential location selection [34]. It is interesting to note that residential preferences for walking and access to transportation among college-educated youth living in single-person households are strong when considering the location index and this index [35]. This is suggested by the classical theory of urban economics, which assumes that households will pay higher prices for more accessible land to save on commuting costs. Proximity-based index by low-carbon mobility programs helps to create favourable conditions for activity and carbon-neutral mobility [36].

5.6 Travel time-based criterion

The relationship of this criterion with access has been strongly positive and significant, and other access variables of the location-based index sub-group are estimated to be marginally significant. The results also show that in places where, in particular, access to public and automatic transportation has a significant impact on choosing a residential location, commuting time (automatically) had the greatest impact on the choice. Travel time calculations can also include a mode index, often based on available data from transportation models and programs. Many empirical studies show that travel time uncertainty has significant effects on people's travel planning [37], [38]. For example, Abdel-Aty et al. [39] determined that this issue is either the most important or the second most influential factor in travellers' decision-making. Also, previous studies have shown that people exposed to travel time uncertainty tend to avoid the risk of being late and consider additional travel time to ensure the probability of arriving on time to participate in activities. Another research in China shows that travel time should be included in location-based access because it may lead to better and more reliable decision-making. Therefore, urban planning can adopt better policies for different social groups with travel reliability [40].

6 TRANSPORT-BASED CRITERION

The transportation-based criterion describes the transportation system, expressed as its usefulness for covering the distance between the origin and the destination using a specific transportation method. Some studies consider transportation-based accessibility based on accessibility to transportation stations as a criterion for accessibility, although destinations are not included in it [41]. This criterion can accommodate the other subcategories and refer to them as well: measures based on infrastructure and measures based on the environmental dimension.

6.1 Measures based on infrastructure

Infrastructure-based measures are transport-based measures that provide insight into the performance or service level of transport infrastructure (such as average travel speed on the road network). Considering that it is related to the transportation component, it hardly communicates with other indicators because it focuses on the quality of the transportation network and does not provide insight into the levels of access experienced between different areas or household groups in the same area. On the other hand, compact urban development with mixed-use, following low-carbon transportation systems is considered a vanguard for mobility. Such conditions allow travellers to choose sustainable modes of transportation or minimize travel time [13], [42]; however, it is of secondary importance.

6.2 Measures based on the environmental dimension

In terms of the environmental dimension, attention has been focused on the difference between transportation methods in terms of energy consumption and their external effects [43]. This criterion also emphasizes the degree of dependence of an urban area or a part of it on a type of transportation method based on the accessibility. Considering the side effects related to the transportation methods that have placed car accessibility significantly higher than public access, it is possible to define a car-dependent area. It is expected that when the share of car and medal systems in an area is high, the negative side effects related to transportation are practical. Of course, this indicator can be reduced to the lowest level by improving accessibility with alternative transportation methods. Therefore, the comparison between access to the car versus public transportation at the level of a region can be discussed and analysed. An article shows that reducing greenhouse gas emissions related to transportation becomes a kind of access tool and considers the main implementation barriers to be the separation of land use and transportation planning, which describes the relevance of an integrated concept [44].

7 TIME-BASED CRITERION

This index reflects time constraints, such as the availability of opportunities at different times of the day and the time available to individuals to engage in specific activities (such as work, study, or recreation). Automatic access in this index indicates that delays caused by It include congestion, while for the transit-based index, travel time is used as an input to calculate accessibility. Other time components such as arrival and departure time, waiting time and total travel time can be included in this index using multiple estimates. Therefore, looking at the literature, we can see that they calculate the time at the level of the neighbourhood or region. This review is sufficient for automatic access because a few minutes of additional driving time in the neighbourhood is not necessary for access, but the walking time to the transportation station is for access. Public transport is critical. Since the total transit travel time is also strongly related to the access time of the exit as well as the assumptions related to the waiting time and the number of transfers between the transit lines that are allowed, therefore the accurate assessment by this indicator gives a more realistic picture of the access gaps. It will be shown between families without cars versus people who own cars.

8 CRITERION BASED ON THE INDIVIDUAL COMPONENT AND THE SOCIAL DIMENSION

This index includes access at the individual level, taking into account personal facilities and limitations (the number of activities that a person can participate in at a given time) and socio-economic characteristics. According to research, people have different perceptions of access levels based on individual indicators, and as a priority or limitation, different people show different levels of willingness to travel in order to achieve opportunities, so according to this access index, the concept becomes relative, which can lead to biased results when the absolute parameter of well-being and sustainability is considered.

This indicator is highly dependent on input data and, therefore, difficult to apply at the level of an urban area. The relationship between this index and social ideals has also stimulated conceptual and experimental research in recent years, which contributes to our progress and understanding of the mechanisms and has an impact on examining access, the access status of disadvantaged populations, and the consequences of limited access. At the same time, this index is taken into consideration that the social justice dimension of sustainable development draws attention to the distribution of benefits and burdens of society members, which starting from transportation as a derived need and access as an indicator of the level of participation of all groups in activities. Which are considered normal for sociologists, such as access to employment and essential services.

The study examines the factors affecting the satisfaction of pedestrians in accessing the metro station, taking into account the socio-economic characteristics of the pedestrians and the built environment factors, and shows that under this criterion, access significantly affects the satisfaction of the pedestrians [11]. This index is sometimes included with population classification in location studies; based on this, age group or socio-economic characteristics are considered based on the classification of destinations [45].

Another study examines physical access to food and stores using this index. In this research, the elderly is considered as the target group because this index is significant for participation in social life in the respective societies [46], and on the other hand, the age of people is considered as a key branch in this index, while attention has attracted a limited. In addition, the growing rate of the population, especially the elderly population in developed countries, has raised concerns about access to other prominent places. In a study of bus station accessibility in Melbourne, Australia, Delbosc and Currie [47] found that the low-income population had the lowest level of accessibility after the elderly. This issue becomes important since senior citizens use public and non-motorized transportation methods more than others to cover their daily needs and are dependent on them [48].

A study in Granada, Spain, also shows that the preference to access retail on foot, despite the presence of numerous stores throughout the city, was related to age and social and personal status. Of course, it has also been stated that elderly people with vehicles prefer to use their own vehicles for commuting [49].

9 CONCLUSION

Access refers to people's ability to reach destinations, activities and services, and their quality will have direct and indirect effects. Improving access and its value is often considered as the goal of transportation policy and planning and as one of the channels for improving sustainability, and optimal planning in this case also requires a more comprehensive analysis of access, and no single method can Evaluate all access factors. Therefore, the discovery of a new era of the concept of accessibility will have a significant impact on achieving greater sustainability. The selection of superior evaluation techniques will be very effective, and they can be used to select more than one indicator to help solve transportation and urban planning problems and achieve access to a new way with a decision mechanism that is decisive. Therefore, hybrid scenarios that target a sustainable economy and choose a transportation system by reducing the use of private vehicles and increasing accessibility and creating a liveable city that also minimizes environmental impacts are considered. Because improving access can help achieve many economic, social and environmental goals by reducing related costs. On the other hand, many factors are influential in achieving this concept, such as people's transportation needs and abilities, the quality and affordability of transportation options, land use patterns, and the quality of transportation alternatives; Of course, consultation with local authorities and economic activists in this field should not be ignored. Because consultation with local authorities and economic stakeholders is essential for improvement, and the publication and visual display of accessibility criteria can significantly increase understanding and include a wide range of stakeholders. Therefore, it will help to further present this critical challenge.

REFERENCES

[1] Geurs, K.T., Krizek, K.J. & Reggiani, A., Accessibility analysis and transport planning: An introduction. *Accessibility Analysis and Transport Planning: Challenges for Europe and North America*, pp. 1–14, 2012.

[2] Banister, D., The sustainable mobility paradigm. *Transport Policy*, **15**(2), pp. 73–80, 2008. DOI: 10.1016/j.tranpol.2007.10.005.

[3] Foth, N., Manaugh, K. & El-Geneidy, A.M., Towards equitable transit: Examining transit accessibility and social need in Toronto, Canada, 1996–2006. *Journal of Transport Geography*, **29**, pp. 1–10, 2013. DOI: 10.1016/j.jtrangeo.2012.12.008.

[4] Helling, A., Changing intra-metropolitan accessibility in the U.S.: Evidence from Atlanta. *Progress in Planning*, **49**(2), pp. iii–107, 1998.

[5] Silva, C., Accessibility instruments for planning practice in Europe. COST action TU1002. *Esof2012*, 11–15 Jul., 2012.

[6] Bruins Ma, F., Nij, R. & Rietveld, P., *Serie Research Memoranda Infrastructure and Metropolitan Development in an International Perspective: Survey and Methodological Exploration*, 1990.

[7] Hansen, W., How accessibility shapes land use. *J. Am. Inst. Plan.*, **25**, pp. 73–76, 1959.

[8] Marwal, A. & Silva, E., Literature review of accessibility measures and models used in land use and transportation planning in last 5 years. *Journal of Geographical Sciences*, **32**, pp. 560–584, 2022.

[9] Kelobonye, K., Xia, J.C., Swapan, M.S.H., McCarney, G. & Zhou, H., Drivers of change in urban growth patterns: A transport perspective from Perth, Western Australia. *Urban Science*, **3**(2), p. 40, 2019.

[10] Siddiq, F. & Taylor, B.D., Tools of the trade? Assessing the progress of accessibility measures for planning practice. *Journal of the American Planning Association*, **87**(4), pp. 497–511, 2021. DOI: 10.1080/01944363.2021.1899036.

[11] Bivina, G.R., Gupta, A. & Parida, M., Walk accessibility to metro stations: An analysis based on meso- or micro-scale built environment factors. *Sustainable Cities and Society*, **55**, 102047, 2020. DOI: 10.1016/j.scs.2020.102047.

[12] Wachs, M. & Kumagai, T.G., Physical accessibility as a social indicator. *Socio-Economic Planning Sciences*, **7**(5), pp. 437–456, 1973.

[13] Geurs, K.T. & van Wee, B., Accessibility evaluation of land-use and transport strategies: Review and research directions. *Journal of Transport Geography*, **12**(2), pp. 127–140, 2004.

[14] Allen, B., Liu, D. & Singers, S., Accessibility measures of U.S. metropolitan areas. *Transp. Res. Part B Methodol.*, **278**(6), pp. 439–449, 1993.

[15] Litman, T., Evaluating accessibility for transportation planning: Measuring people's ability to reach desired goods and activities. *Transportation Research*, p. 62, 2017.

[16] Curtis, C., Planning for sustainable accessibility: The implementation challenge. *Transport Policy*, **15**(2), pp. 104–112, 2008.

[17] Bertolini, L., le Clercq, F. & Kapoen, L., Sustainable accessibility: A conceptual framework to integrate transport and land use plan-making. Two test-applications in the Netherlands and a reflection on the way forward. *Transport Policy*, **12**(3), pp. 207–220, 2005.

[18] Martens, K., Justice in transport as justice in accessibility: Applying Walzer's "Spheres of Justice" to the transport sector. *Transportation (Amst)*, **39**(6), pp. 1035–1053, 2012.

[19] Preston, J. & Rajé, F., Accessibility, mobility and transport-related social exclusion. *Journal of Transport Geography*, **15**(3), pp. 151–160, 2007.

[20] Proffitt, D.G., Bartholomew, K., Ewing, R. & Miller, H.J., Accessibility planning in American metropolitan areas: Are we there yet? *Urban Studies*, **56**(1), pp. 167–192, 2019.

[21] Handy, S., Planning for accessibility: In theory and in practice. *Access to Destinations*, eds D.M. Levinson and K.J. Krizek, Emerald Group: Bingley, pp. 131–147, 2005. DOI: 10.1108/9780080460550-007.

[22] Halden, D., The use and abuse of accessibility measures in UK passenger transport planning. *Research in Transportation Business and Management*, **2**, pp. 12–19, 2011. DOI: 10.1016/j.rtbm.2011.05.001.

[23] Boisjoly, G. & El-Geneidy, A.M., How to get there? A critical assessment of accessibility objectives and indicators in metropolitan transportation plans. *Transport Policy*, **55**, pp. 38–50, 2017. DOI: 10.1016/j.tranpol.2016.12.011.

[24] Handy, S.L. & Niemeier, D.A., Measuring accessibility: An exploration of issues and alternatives. *Environment and Planning A*, **29**, pp. 1175–1195, 1997.

[25] Papa, E. & Coppola, P., Gravity-based accessibility measures for integrated transport-land use planning (GraBAM). Accessibility Instruments for Planning Practice Report 1 of COST Action TU1002, 2012.

[26] Palacios, M.S. & El-Geneidy, A., Cumulative versus gravity-based accessibility measures: Which one to use? *Findings*, 2022. DOI: 10.32866/001c.32444.

[27] Merlin, L.A. & Hu, L., Does competition matter in measures of job accessibility? Explaining employment in Los Angeles. *Journal of Transport Geography*, **64**, pp. 77–88, 2017. 10.1016/j.jtrangeo.2017.08.009.

[28] Kelobonye, K., Zhou, H., McCarney, G. & Xia, J., Measuring the accessibility and spatial equity of urban services under competition using the cumulative opportunities measure. *Journal of Transport Geography*, **85**, 102706, 2020. DOI: 10.1016/j.jtrangeo.2020.102706.

[29] Nasri, A. & Zhang, L., Multi-level urban form and commuting mode share in rail station areas across the United States: A seemingly unrelated regression approach. *Transport Policy*, **81**, pp. 311–319, 2019. DOI: 10.1016/j.tranpol.2018.05.011.

[30] Ewing, R. & Cervero, R., Travel and the built environment. *Journal of the American Planning Association*, **76**(3), pp. 265–294, 2010.

[31] Cervero, R., Sarmiento, O.L., Jacoby, E., Gomez, L.F. & Neiman, A., Influences of built environments on walking and cycling: Lessons from Bogotá. *International Journal of Sustainable Transportation*, **3**(4), pp. 203–226, 2009.

[32] Næss, P., Strand, A., Wolday, F. & Stefansdottir, H., Residential location, commuting and non-work travel in two urban areas of different size and with different center structures. *Progress in Planning*, **128**, pp. 1–36, 2019.

[33] Kaufman, B., Leung, A. & Burke, M., Evaluating demand responsive transit services using a density-based trip rate metric. *Journal of Transport and Land Use*, **14**(1), pp. 499–519, 2021.

[34] Guo, J.Y. & Bhat, C.R., Operationalizing the concept of neighborhood: Application to residential location choice analysis. *Journal of Transport Geography*, **15**(1), pp. 31–45, 2007.

[35] Myers, D. & Gearin, E., Current preferences and future demand for denser residential environments. *Housing Policy Debate*, **12**(4), pp. 633–659, 2001.

[36] Pajares, E., Büttner, B., Jehle, U., Nichols, A. & Wulfhorst, G., Accessibility by proximity: Addressing the lack of interactive accessibility instruments for active mobility. *Journal of Transport Geography*, **93**, 103080, 2021. DOI: 10.1016/j.jtrangeo.2021.103080.

[37] Carrion, C. & Levinson, D., Value of travel time reliability: A review of current evidence. *Transportation Research Part A: Policy and Practice*, **46**(4), pp. 720–741, 2012. DOI: 10.1016/j.tra.2012.01.003.

[38] Taylor, M.A.P., Travel through time: The story of research on travel time reliability. *Transportmetrica B*, **1**(3), pp. 174–194, 2013.

[39] Abdel-Aty, M.A., Kitamura, R. & Jovanis, P.P., Investigating effect of travel time variability on route choice using repeated-measurement stated preference data. *Transportation Research Record*, **1493**, pp. 39–45, 1995.

[40] Chen, B.Y. et al., Measuring place-based accessibility under travel time uncertainty. *International Journal of Geographical Information Science*, **31**(4), pp. 783–804, 2017. DOI: 10.1080/13658816.2016.1238919.

[41] Levine, J., Grengs, J. & Shen, Q., Metropolitan accessibility and transportation sustainability: Comparative indicators for policy reform final report from EPA Agreement Number: RD-833334901-0 and FHWA Cooperative Agreement.

[42] Banister, D., Cities, mobility and climate change. *Journal of Transport Geography*, **19**(6), pp. 1538–1546, 2011. DOI: 10.1016/j.jtrangeo.2011.03.009.

[43] Feitelson, E., Introducing environmental equity dimensions into the sustainable transport discourse: issues and pitfalls. www.elsevier.com/locate/trd.

[44] Kinigadner, J. & Büttner, B., How accessibility instruments contribute to a low carbon mobility transition: Lessons from planning practice in the Munich region. *Transport Policy*, **111**, pp. 157–167, 2021.

[45] Fan, Y., Guthrie, A. & Levinson, D., Impact of light-rail implementation on labor market accessibility: A transportation equity perspective. *Journal of Transport and Land Use*, **5**(3), pp. 28–39, 2012.

[46] Lucas, K., van Wee, B. & Maat, K., A method to evaluate equitable accessibility: Combining ethical theories and accessibility-based approaches. *Transportation (Amst)*, **43**(3), pp. 473–490, 2016. DOI: 10.1007/s11116-015-9585-2.

[47] Delbosc, A. & Currie, G., Using Lorenz curves to assess public transport equity. *Journal of Transport Geography*, **19**(6), pp. 1252–1259, 2011. DOI: 10.1016/j.jtrangeo.2011.02.008.

[48] Paez, A., Mercado, R.G., Farber, S., Morency, C. & Roorda, M., Accessibility to health care facilities in Montreal Island: An application of relative accessibility indicators from the perspective of senior and non-senior residents. *International Journal of Health Geographics*, **9**, pp. 1–15, 2010.

[49] Ariza-Álvarez, A., Arranz-López, A. & Soria-Lara, J.A., Comparing walking accessibility variations between groceries and other retail activities for seniors. *Research in Transportation Economics*, **87**, 100745, 2021.

LANDUSE PLANNING FOR ELECTRIC VEHICLE INFRASTRUCTURE

GARGI GHOSH
Sky Group, India

ABSTRACT

India is committed to the Paris Agreement by making efforts to reduce the emissions intensity of its GDP by 20–25% from its 2005 levels by 2020 and by 33–35% by 2030. As a part of this commitment, India intends to adopt 30% electric vehicles (EVs) in its vehicle population by 2030. In pursuant to the mission, the government has formulated the Electric Vehicle Policy, Model Building Guidelines and other related policies and guidelines. Various financial tools and incentives have also been declared. Large scale adoption of EVs would require provision of various infrastructures for the EV users. One of the major requirements is the availability of accessible charging infrastructure. Providing accessible charging infrastructure for EVs within the urban landscape would pose various locational, financial and technical challenges. Needless to say, there is a need for identifying spatial and landuse planning strategies to accommodate and sustainably plan for EV infrastructure. There has been a good initiative of work on locational models, plans, user perspective and business models for charging stations. However, works on landuse planning for charging infrastructure is still in its nascent stage. The aims of the paper are to (a) explore the necessity of deriving a landuse planning framework for EV charging infrastructure; and (b) to identify a replicable model for spatial planning for EV charging infrastructure for Indian cities. The paper reviews the various applicable guidelines for EV infrastructure and best practices, locational theories, zoning and building regulations and arrives at a framework for landuse planning for EV infrastructure. The case area for the study is Bangalore city.

Keywords: electric vehicles, charging infrastructure, India, Bangalore, landuse planning, AHP.

1 INTRODUCTION

India has committed to five key elements of the Climate Plan at the 26th session of the Conference of the Parties (COP26) to the United Nations Framework Convention on Climate Change (UNFCCC) in continuation with the Paris accord of 2015. The five key elements of the action plan are (i) reach 500GW non-fossil energy capacity by 2030; (ii) 50% of its energy requirements from renewable energy by 2030; (iii) reduction of total projected carbon emissions by one billion tonnes from now to 2030; (iv) reduction of the carbon intensity of the economy by 45% by 2030, over 2005 levels; and (v) achieving the target of net zero emissions by 2070 [1]. The government of India has taken multiple strategic policy initiatives in the energy and transport sector to steer the nation to the path of achieving the climate plan.

National Electric Mobility Mission Plan 2020 is one of the key pillars of the plan promoting electric vehicle (EV) adoption in the market. The mission plan is accompanied by several operational, finance and technical schemes and guidelines to expedite EV adoption. EV markets and user characteristics worldwide has shown that EV adoption requires transcending the challenges on EV charging infrastructure (EVCI). The EV market in India is at a nascent stage and EVCI has been sparsely developed. Standardised models and guidelines for development and operation of EVCI in the Indian context are lacking. This paper aims to (a) explore the necessity of deriving a landuse planning framework for EV charging infrastructure; and (b) to identify a replicable model for landuse planning for EV charging infrastructure for Indian cities.

WIT Transactions on The Built Environment, Vol 212, © 2022 WIT Press
www.witpress.com, ISSN 1743-3509 (on-line)
doi:10.2495/UMT220061

2 LITERATURE REVIEW

Road transport is the largest consumer of oil in India and the third largest sector in terms of CO_2 emission, amounting to 1.69 tonnes of CO_2 emission per capita in 2019 [2]. Hence India has taken a conscious decision to move from internal combustion engines to battery electric vehicles. Electric Vehicles (EV) were slow in uptake in the market. However with advancement in battery technology, the potential of EV penetration in the vehicle market is promising. Efficient energy storage and charging of EV batteries have been at the epicentre of discussion in the EV ecosystem. EV batteries chargers are of two types – slow chargers (for home charging) and fast chargers (for public places) [3]. It has been widely studied that availability of charging infrastructure is a major concern for adoption of electric vehicles [4]. Sustainable EV charging business models on the other hand require adequate market demand. Locational models are hence important to resolve the conflicting situation [5]. Norway has one of the largest EV fleet in the world. Studies from the Norwegian EV market show that with growing EV penetration demand large scale public charging infrastructure in addition to home based charging system. A well spread public charging network has also shown increased propensity to encourage complete shift to EV in the household [6].

Research on EV and related infrastructure has gained momentum in the last decade. In countries with high EV penetration, the coverage of research is widespread. Research is found in the five main areas, (a) Needs assessment and gaps; (b) energy and finance; (c) sital design standards; (d) user experience; and (e) locational models. In the Indian context literature is found in areas of grid integration, operational models and locational models. Literature with respect to landuse and EV charging infrastructure is scarce. For the purpose of this paper, literature on locational models at global scale has been studied along with papers on policies, guidelines and literature on EV available in Indian context

2.1 Locational models

Published literature portrays a large array of charging station locational models. Majority of this research are within the last decade. The studies are focused on either economic impacts or power grid impacts. Among studies considering economic aspects, research on locational models is primarily of two types, (a) theoretical models; and (b) empirical applications. Theoretical works have concentrated on deriving mathematical models to assess locations for EV infrastructure with simulated data whereas empirical models focus on real spaces with spatial characters often with georeferenced data and provide proof of concept for various mathematical locational models. EV charging infrastructure site selection is a multiple criteria decision making (MCDM) problem as it is determined by several factors often contradictory [7]. Literature shows that AHP and fuzzy AHP have been used to determine the weights of criteria. PROMETHEE method has been used with four main and seven sub-criteria in a case study in Beijing to derive geographical locations of EV charging sites [8]. In addition to mentioned MCDM methods, fuzzy TOPSIS method [9], PROMETHEE and ANP [10] approaches, fuzzy Delphi and GRA-VIKOR combination [11] have also been seen in research as methods to arrive at results via MCDM in various case studies conducted in Beijing. GIS is a prominent tool in studies that work with georeferenced data. GIS based MCDM [12]–[14] methods have also been used for optimal site selection in Ankara, Istanbul and Spain.

Various evaluations attributes are found in literature discussed above which is used for the MCDM process for deriving site suitability. The evaluation attributes are compiled under the following heads – traffic factors, economic factors, social factors and influencing factors and presented in Table 1.

Table 1: Evaluation attributes compiled from literature.

Environmental	Economic	Service capacity	Societal
Destruction level on vegetation and water	Construction cost	Traffic convenience	Attitude of local residents
Waste discharge	Annual operation and maintenance cost	Daily service capacity	Local government support
Air pollutant reduction	Investment payoff period	Car ownership	
Adverse impact due to sound, dust, etc.		Proximity to taxi stand	
Distance to water resources		Traffic density	
Distance to landslide risk		Proximity to interchange stations	
Slope		Road network	
Topography		Proximity to LRT stations	
		Distance between CS and user	
		Distance between CS and destination	

Reliable data is the key to achieving realistic outputs. Data used for the site selection studies are either statistical data availed from Census or travel data for EV and travel data of conventional fuel vehicles, surveys or simulation data. Statistical data is found to be useful in user-oriented models. Residential population, income, socio-economic statistics have been used for assessing demand density for EV charging stations. Travel data from GPS tracking of EV or conventional vehicles have been used for predicting driving tendency to analyse EV charging locations. Survey data with driving behaviour, origin and destination as well as temporal information has also been used for charging station locational analysis.

The available literature, however, does not look into EV charging infrastructure from the point of view of landuse planning policy or zoning regulations. Hence, suitability criteria focused on landuse zoning and analysis with landuse focused data sets are scarce in literature.

2.2 Indian EV infrastructure planning scenario

There are 285 million registered vehicles in India [15]. As of 2021, the share of EVs in the market is less than 1%. The government of India has targeted 30% EV penetration by 2030. To achieve that a sales penetration of 80% for two and three-wheelers, 50% for four wheelers, and 40% for buses by 2030 is required [16]. Several steps have been taken by the government for facilitating and encouraging the market. Financial scheme like FAME I and II for financial nudging of manufacturing, technical specifications for charging system, guidelines for charging infrastructure, amendments to Model Building byelaws, planning guidelines (URDPFI) and battery swapping policy have been published. A planning guideline of at least one charging station in a 3 km × 3 km grid is prescribed, one charging station per 25 km along highways is recommended while at a regional level one charging station in a grid of 10 km × 10 km is desired in designated industrial parks, SEZs and estates [17]. 20% of available parking space will be considered for EV parking and related charging infrastructure in residences, points of work and commercial places [18]. EV charging stations are permitted in central district, central sub district, district centres, community centres, residential and industrial landuse zone in urban areas, along National Highways, identified growth node villages and along major roads. Minimum sital area of 13.5 m × 5.5 m is recommended for charging stations whereas minimum area if 5.5 m × 2.75 m is recommended for battery swapping stations [19].

A review of the guidelines on EVCI in India reveals that EVCI is being perceived in the same framework as fuel stations. However, long charging time requirement mandates EVCI to be treated in a distinct manner. EV vehicle adoption is anticipated at a higher rate in urban areas compared to rural areas. Market reports predict higher EV adoption in 2-wheeler, 3-wheeler, intra-city buses and light goods vehicles based on the availability of charging infrastructure. While 2-wheeler and 3-wheeler category vehicles can be charged at home and city buses and goods vehicles can be charged at the parking lots, no such ready options exist for 4-wheelers making its adoption and in turn achievability of India's EV target arduous [20]. Hence city based EVCI will be a major requirement. In dense urban setting of Indian cities EVCI will have to compete will lack of vacant space and steep real estate price. This warrants a relook into the landuse planning and zoning regulations in cities.

3 METHODOLOGY

The present paper attempts to derive a landuse planning framework for EV charging infrastructure in two steps. Firstly suitable EV charging sites have been identified using multi criteria decision making methods and GIS using empirical data. The resultant output provides the basis of evaluating landuse policies with respect to sites suitable for EV infrastructure. Thereafter the present landuse zoning and regulation have been studied to derive an ideal landuse planning framework for EV charging infrastructure.

A list of decision criteria was compiled from those found in literature and presented in Table 1. Experts in the field of spatial planning, energy and GIS were interviewed to shortlist the critical criteria and assign prioritising scores as per Table 2. The priority scores were used for pairwise comparison in AHP. 10 sets of interview data were collected from this study. "Proximity to roads", "proximity to residential areas", "proximity to work centres and commercial centres", "proximity to interchanges" and "availability of open space" were the five criteria shortlisted for the present paper. "Proximity to road" represented accessibility and a higher accessibility was more suitable for locating EVCI. "Proximity to residential areas" and "proximity to work centres and commercial centres" represented demand, a higher proximity to residential areas represented higher demand for the EVCI which in turn represented higher financial feasibility. "Proximity to interchanges" like metro stations, bus stations represented demand. "Availability of open space" represented open space inside a property that could be used to set up charging stations.

Table 2: Rating used for prioritising criteria.

Sl	Preferences	Rating
1	Equally important	1
2	Moderately important	3
3	Strongly important	5
4	Very important	7
5	Extremely important	9

AHP was used for deriving the weights of the criteria, as shown in Table 3. Pair wise comparison of the five shortlisted criteria was conducted as per the relative scores from expert interview. Standardised matrix was derived and thereafter the criterion weights and consistency ratio (CR) analysis was performed to verify the accuracy of the AHP process. The sum total of the weights of the criteria is one. The CR of the AHP analysis was found to 0.034 which was well within the acceptable limits.

Table 3: Weights of criteria as derived from AHP.

Sl.	Criteria	Weights
1	Proximity to roads	0.167
2	Proximity to residential areas	0.167
3	Proximity to work centres and commercial centres	0.167
4	Proximity to interchanges	0.079
5	Availability of open space	0.42
		1.00

The data for GIS analysis corresponding to the above parameters were availed from public domain. The GIS layers pertaining to roads, landuse and interchange locations were suitably updated to reflect current situation in the study area. Open space data was available by deducting building footprint area from property area. Since the accuracy of this data directly affects the reliability of the study, great care was taken during the data transfer. Euclidean distance analysis was adopted for roads, work centres, residential and interchanges areas. Open space data was categorised into "less than 100 m^2" and "more than 100 m^2" categories and thereafter normalised. Weighted overlay analysis was performed on the normalised raster layers with the weights of the criteria obtained from AHP.

The study area is Bangalore city. Bangalore is the third largest metropolis in India after Mumbai and New Delhi. It has the second highest EV adoption rate in India at present. The city has a population of about 10 million. It can be considered as a representative urban area in the country.

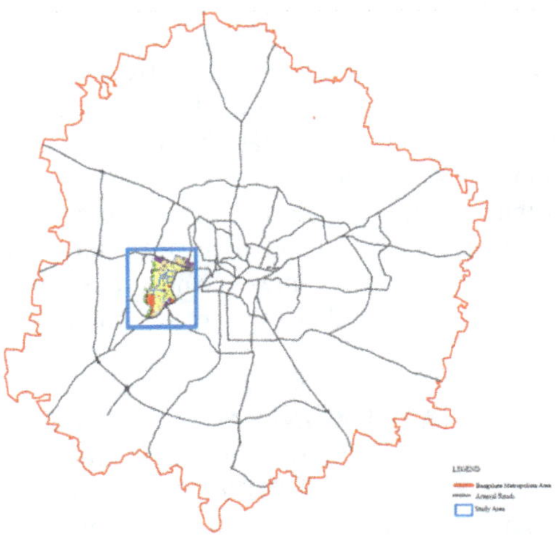

Figure 1: Location of study area in Bangalore. *(Source: Bangalore Revised Master Plan 2015.)*

A part of the city's western area measuring about 1.65 km^2 was taken up for the study, as shown in Fig. 1. The existing landuse zoning area is characterised by middle to upper middle class residential areas interspersed with neighbourhood level markets, zonal markets, well distributed healthcare and educational facilities, parks and government offices, as shown in Fig. 2. The area is connected to the rest of the city via road and metro rail. There are no major variations in slope, forest, water bodies or other protected areas. The urban fabric of the area is dense and representative of the urban fabric of the city.

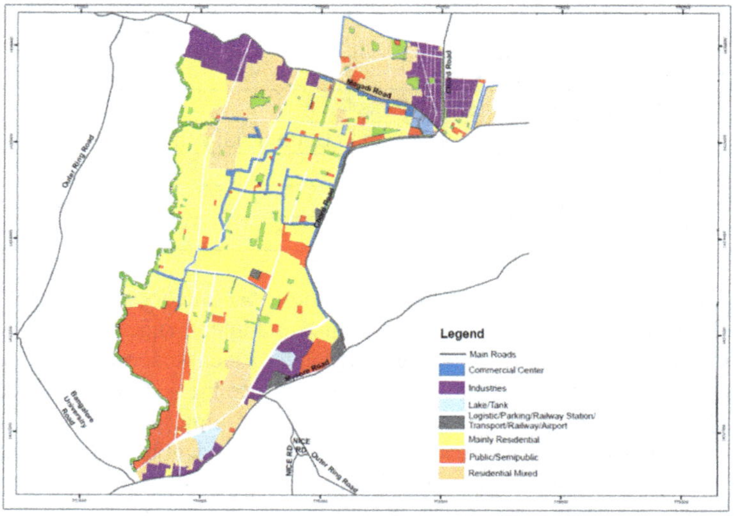

Figure 2: Landuse zoning in study area. *(Source: Bangalore Revised Master Plan 2015.)*

4 RESULTS

The weighted overlay analysis used the spatial layers of the five factors used for the multi criteria decision making model. The suitability output of the individual layers is shown in Fig. 3.

The weighted overlay analysis using the weights derived from AHP generated a graded suitability map of the study area, as shown in Fig. 4. The output was further reclassified to generated sites scoring 0.5 and above in the suitability score to generate "more suitable sites" for EVCI. The more suitable site locations were then studies with respect to their present zoning. It was revealed that the suitable site locations were at present zoned under parks, transportation use (bus depots, metro station premises) and public and semi-public use (hospitals, schools, government office premises) and industrial use zones. The suitable locations did not coincide with residential or commercial landuse.

5 LANDUSE PLANNING WITH RESPECT TO EV CHARGING INFRASTRUCTURE

EVCI is a relatively new concept in India with an introductory set of guidelines related to building bye laws as well as zoning. The URDPFI (urban and regional development plan formulation and implementation) guidelines serve as a model for urban planning published by the Ministry of Housing and Urban Development, government of India. The guideline has however no legal bearing upon the cities. Landuse zoning as well as development regulations

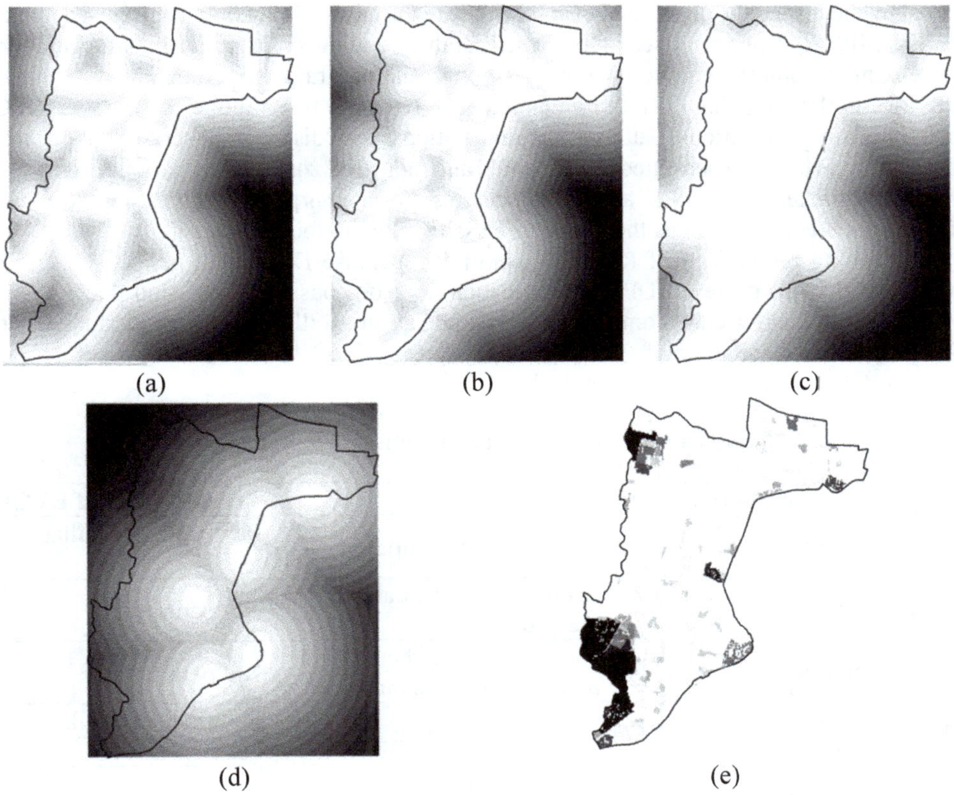

Figure 3: GIS output of each criteria layer. (a) Proximity to major roads; (b) Proximity to work centres; (c) Proximity to residential areas; (d) Proximity to interchanges; and (d) Availability of vacant land.

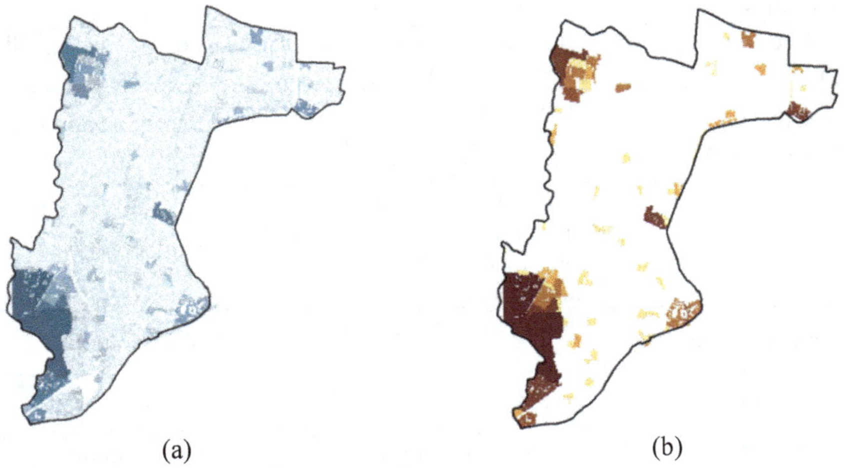

Figure 4: Suitable sites EV charging stations. (a) Suitability map; and (b) More suitable locations.

are set by the respective development authorities. In the case of Bangalore the Revised Master Plan (RMP) 2015 with its set of proposed landuse zoning maps and zoning regulations (development control regulations) form the legal regulatory framework for all landuse related activities in the city. The URDPFI amendment in response to EV policies classifies EVCI including charging stations, battery swapping stations as a "distribution service" at par with fuel stations. These are permitted in residential and industrial zones, along national highways, freight complexes, city centres and district centres. For the purpose of evaluating the zoning and development regulations, the URDPFI classification has been considered.

The zoning regulations of RMP 2015 do not recognise EVCI as a permissible use at present. However, if the URDPFI classification is to be considered, we find that EVCI is permitted as a main use in "commercial axes", and as an ancillary use (permitted in 20% of site area) in "residential main", "commercial business zones", "public and semi-public" and "transportation zones" as seen in Table 4.

Table 4: Landuse zones and permissibility of EVCI.

Sl	Landuse category	Description	Permissibility of EVCI
1	Residential main	Area of mainly residential in use	Permitted as ancillary use
2	Residential mixed	Area of mixed residential and commercial use	Not permitted
3	Commercial central	Commercial areas in the core city mainly wholesale markets	Not permitted
4	Commercial business	Central business district	Permitted as ancillary use
5	Mutation corridor	One property deep areas along the radial arterial roads	Not permitted
6	Commercial axes	Commercial areas within residential zones	Permitted as main use
7	Industrial general	All industries except IT and ITES	Not permitted
8	Industrial hi-tech	IT and ITES industry	Not permitted
9	Public and semi public	Education, healthcare and government buildings	Permitted as ancillary use
10	Traffic and transportation	Premises of transport infrastructure – bus depot, metro station, parking lots	Permitted as ancillary use
11	Public utility	Premises of water supply, sewerage, electricity, power solid waste management infrastructure	Not permitted
12	Park and open space	Parks, playgrounds, stadiums, pool, libraries, cemeteries	Not permitted
13	Agricultural use	Agricultural areas with no developable zone	Not permitted
14	Unclassified	Defence establishments	Not permitted

5.1 Zoning of electric vehicle charging infrastructure

EVCI has been categorised similar to fuel stations at present. While fuel stations present very risk of life and property destruction [21] EVCI hazards from EVCI connected with smart grid and communication server has proven to be comparatively low [22]. Hence there is a need to distinguish EVCI from fuel stations. Fuel stations at their present grid density cater to the market demand. Charging station requirement will be much higher owing to the longer time requirement per vehicle. It calls for a necessity to recognise EVCI in terms of zoning so that large numbers of such structures can be integrated into urban landuse with due considerations of safety.

5.2 Permissibility of electric vehicle charging infrastructure in various landuse zones

The results of the site suitability analysis represent the feasible locations of setting EVCI for public use. As seen in the previous section, the feasible sites also coincided with parks and public and semi-public use and transportation zones. Parks are not considered feasible sites as they are not developable lands.

Public and semi-public areas are work centres and with exceptions to private schools and hospitals, are often located on government land. These premises tend to be large with low density development. The situation is similar is most of the Indian cities. Space availability and public ownership would provide lucrative market opportunity for use as EVCI for 2-wheelers as well as 4-wheelers. The zoning regulations at present does not permit EVCI in Public and semi-public lands. If considered similar to fuel stations, it is permitted as an ancillary use, i.e. up to 20% of the site area. However, based on the preceding discussions in the paper, it would be helpful to amend the regulations to ensure EVCI supply in the urban market.

Traffic and transportation zones are premised of railway stations, metro stations, bus depots and parking lots. These are located close to work centres or residential neighbourhoods as major transport infrastructure are planned to connect work/business centres to residential centres. These also have large premises and are almost always on government owned lands. These provide opportunity for setting up for charging stations for 4-wheelers as well as buses and goods vehicles where vehicles can be left for long hours overnight or during the day. The conditions would be similar in most Indian cities. The zoning regulations again do not recognise EVCI and if considered similar to fuel stations, is permitted as ancillary use only. It is imperative that the zoning permits EVCI on these zones as a main use to ensure market supply for public charging infrastructure.

Considering the approach of the URDPFI, EVCI should also be a permitted use in industrial and commercial zones.

6 LIMITATIONS OF THE STUDY

The study area selected for the study is a portion of the city. It is meant to be representative of the urban fabric of Bangalore while facilitating ease of data handling and ensuring realistic results. The data used for the study was also current and great care was used for data transfer to arrive at accurate results. However, study conducted at the city scale would add several infrastructure and landuse parameters to the study which may add different dimensions to the outputs. Such results with diverse and large datasets would add value to policy decisions for landuse and zoning. Hence study with city scale data would be more pertinent for study for policy decisions.

The present study focuses on parameters related to demand, accessibility and space availability based on expert's opinion and availability of reliable data. Environmental parameters may be added for the AHP and subsequent GIS analysis to receive environmentally sensitive outputs.

7 CONCLUSION

The paper presents a model of landuse planning with respect to EV charging infrastructure. The study area is a part of Bangalore city, which is representative of urban fabric of typical Indian cities. The study area hence is very relevant for the study considering high penetration of EVs expected in Indian cities. Accurate and updated datasets have been used for the analysis. The process used for analysis of data is well established in literature. The study is quantitative as well as empirical in nature and presents statistically relevant and realistic outputs.

Policy decisions related to landuse planning is often driven by land market trends. Integrating the transport network and travel pattern of the city with the urbanisation and land market trends is the key to sustainability. This study provides a unique methodology to arrive at landuse policy decisions related to EV charging infrastructure. Bangalore can be considered representative of general Indian city conditions and the outputs can be considered realistic for other Indian cities as well. The process is hence replicable for other cities and study areas.

Landuse policy sensitive to EV needs are highly recommended at this stage of the EV adoption curve in India. EV adoption rates can be increased only by increasing the EV charging infrastructure supply in the market. However, the urban land market would pose distinct challenges as highlighted in the paper. Hence innovative approaches are required to use the urban land to ensure the supply. The present paper uses on ground data to reveal suitable sites and thereafter proposes policy changes for utilisation of the same for EV charging infrastructure.

ACKNOWLEDGEMENTS

I would like to express my gratitude to Sky Group for supporting and funding the research. I also convey a huge vote of thanks to my peers in research and practice for advice and inputs and to my family for their unconditional support.

REFERENCES

[1] Ministry of Environment, Forest and Climate Change, Government of India, Press Release, 3 Feb. 2022.

[2] International Energy Agency, Data and statistics, 2019. https://www.iea.org/data-and-statistics.

[3] Khan, W., Ahmad, F., Ahmad, A., Mohamman Saad, A. & Ahuja, A., Feasibility analysis of electric vehicle charging in India, ISGF, 2017.

[4] Beckers, T., Gizzi, F., Kreft, T. & Hildebrandt, J., Efficient delivery of the (public accessible) charging infrastructure for electric mobility in Germany, TU Berlin, 2015.

[5] Pagany, R., Ramirez Camargo, L. & Dorner, W., A review of spatial localization methodologies for the electric vehicle charging infrastructure. *International Journal of Sustainable Transportation*, **13**(6), pp. 433–449, 2019.

[6] Lorentzen, E., Haugneland, P., Bu, C. & Hauge, E., Charging infrastructure experiences in Norway: The worlds most advanced EV market. *EVS30 Symposium*, 2017.

[7] Tang, D. & Wang, P., Dynamic electric vehicle charging load modelling: From perspective of transportation. Presented at the *2013 4th IEEE/PES Innovative Smart Grid Technologies Europe*, ISGT Europe, 2013.
[8] Wu, Y. & Yang, M., Optimal site selection of electric vehicle charging stations based on a cloud model and the PROMETHEE method. MDPI, 2016.
[9] Zhu, Z.-H., Gao, Z.-Y., Zheng, J.-F. & Du, H.-M., Charging station location problem of plug-in electric vehicles. *Journal of Transport Geography*, 2016.
[10] Wang, X., Yuen, C., Hassan, N., An, N. & Wu, W., Electric vehicle charging station placement for urban public bus systems. *IEEE Transactions on Intelligent Transportation Systems*, 2016.
[11] Zhao, H. & Li, N., Optimal siting of charging stations for electric vehicles based on fuzzy delphi and hybrid multi-criteria decision making approaches from an extended sustainability perspective. *Energies*, **9**(4), 2016.
[12] Ahn, Y. & Yeo, H., An analytical planning model to estimate the optimal density of charging stations for electric vehicles. *PLoS One*, **10**, e0141307, 2015. DOI: journal.pone.0141307.
[13] Erbas¸ M., Kabak, M., Ozceylan, E. & Çetinkaya, C., Optimal siting of electric vehicle charging stations: A GIS-based fuzzy multi-criteria decision analysis. *Energy*, 2018.
[14] Kaya, Ö., Alemdar, K., Atalay, A., Çodur, M. & Tortum, A., Electric car sharing stations site selection from the perspective of sustainability: A GIS-based multi-criteria decision making approach. *Sustainable Energy Technologies and Assessments*, 2022.
[15] Ministry of Road Transport and Highways (MoRTH) Government of India, 2022. https://vahan.parivahan.gov.in/.
[16] GIZ, Federal Ministry of Environment, Nature Conservation and Nuclear Safety of Federal Republic of Germany, NITI Aayog, Status quo analysis of various segments of electric mobility and low carbon passenger road transport in India, 2021.
[17] Ministry of Power, Government of India, Charging infrastructure for electric vehicles, the revised consolidated guidelines and standards, 2022.
[18] Town and Country Planning Organisation, Ministry of Housing and Urban Affairs, Amendments to Model Building Byelaws, 2019.
[19] Town and Country Planning Organisation, Ministry of Housing and Urban Affairs, Amendments to urban and regional development plan formulation and implementation (URDPFI) guidelines for electric vehicle charging infrastructure, 2019.
[20] KPMG, Shifting gears: The evolving electric vehicle landscape in India, 2020.
[21] Ahmed, M.M., Khamidi, M.F., Kutty, S.R.M. & Mohd Shariff, A., Analysis of fuel stations hazards by using risk assessment criteria. *International Conference on Environment*, 2010.
[22] Mitol, M., Dehghanian, P., Wang, S. & Wang, B., Electrical safety considerations in large-scale electric vehicle charging stations. *IEEE Transactions On Industry Applications*, **55**(6), 2019.

URBAN TRANSPORT INFRASTRUCTURE FOR PEOPLE WITH DISABILITIES IN THE CITY OF TSHWANE, SOUTH AFRICA: AN OBSERVATION AUDIT

BABRA DURI* & ROSE LUKE†
Department of Transport and Supply Chain Management, University of Johannesburg, South Africa

ABSTRACT
People with disabilities often find it difficult to access public transport. The purpose of this study was to identify issues with urban transport infrastructure in the City of Tshwane that contribute to inaccessible transportation. An observational checklist was prepared to collect data, and the checklist was based on guidelines and best practices for universally accessible transport. The infrastructure elements considered in this study were pavements, street crossings, street furniture, bus stops, bus station, taxi ranks and train stations. To a large extent, the design of transport infrastructure in the City of Tshwane does not comply with universal design principles and, in areas where there was compliance, there was a distinct lack of consistency. The study results revealed that the urban transport infrastructure in the City of Tshwane is not designed to cater to the needs of people with disability. The results also revealed that existing transport infrastructure lacks maintenance. It is recommended that the city train city planners and built environment professionals on disability building regulations. City planners and built environment professionals should be conscious of the needs of different groups of people when planning and designing city infrastructure.
Keywords: *transport barriers, people with disability, transport infrastructure, built environment, pedestrian environment, universal access.*

1 INTRODUCTION

Urban transport enables city dwellers to access goods, services and opportunities offered by the city. However, the provision of urban transport infrastructure and services is not equal across different groups of transport users [1]. The impact of unequal provision of urban transport infrastructure and services is more pronounced to disadvantaged groups such as people with disability and the elderly. People with disability often encounter numerous barriers in accessing transport infrastructure and services, which may lead to poor quality of life, lack of opportunities, unemployment, poverty [2]–[4]. According to Hernández [1] and Odufuwa [2], transport is one of the essential components of sustainable livelihood which enhances good quality of life. One of the aspirations of the United Nations Sustainable Development Goals (SDG) 11 is that "By 2030, provide access to safe, affordable, accessible and sustainable transport systems for all, improving road safety, notably by expanding public transport, with special attention to the needs of those in vulnerable situations, women, children, persons with disabilities and older persons". South Africa, as one of the countries which ratified the Convention on the Rights of Persons with Disabilities, has the obligation to improve the provision of urban transport infrastructure and services [5].

The South African transport system is shaped by apartheid spatial planning [6]. The legacy of apartheid spatial planning still exists and has a strong effect on the socio-economic activities and arrangement of urban areas [7]. According to Thomas [6] "the provision of safe, accessible, and affordable public transport infrastructure is a fundamental prerequisite for the socio-economic advancement of the South African population". However, some

* *ORCID: http://orcid.org/0000-0003-2977-4897*
† *ORCID: http://orcid.org/0000-0003-1915-6956*

WIT Transactions on The Built Environment, Vol 212, © 2022 WIT Press
www.witpress.com, ISSN 1743-3509 (on-line)
doi:10.2495/UMT220071

people do not have access to good public transport infrastructure in South Africa. In South Africa, "there is no comprehensive disability legislation that deals exclusively with matters of disability or with people with disabilities" [8]. However, there are pieces of legislation that mention people with disability or deal with matters of disability [8].

The City of Tshwane, which is located in Gauteng province of South Africa, shows an increase in population of people living with some form of disability [7]. According to Stats SA [7], there are approximately 800,000 people living with some form of disability in Gauteng province. Little research has focused on the City of Tshwane, yet there has been an increase in people living with some form of disability. It is inevitable that people with disability will experience transportation problems, given the lack of inclusivity in transport provision in the City of Tshwane. According to Boshoff and Marais [9], the City of Tshwane needs to improve accessibility of its central business district (CBD). The purpose of this study was to identify issues with urban transport infrastructure around the Tshwane CBD that contribute to inaccessible transportation. The secondary objectives formulated to address the main aim of the study are: (1) to establish transport barriers encounters by people with disability; and (2) audit the status quo of the transport infrastructure around the Tshwane CBD.

2 TRANSPORT INFRASTRUCTURE

The infrastructure elements of accessibility considered in this study are pavements, street crossings, street furniture, bus stops, bus station, taxi ranks and train stations. People with visual and mobility disability often encounter many barriers in the road and pedestrian environment [2], [3], [10]–[12]. According to Vanderschuren et al. [13], well-designed pavements should have: "sufficient widths; even surfaces with sufficient friction; sufficient lighting; and appropriate placing of street furniture". Ideally the width of pavements should be at least 2,000 mm wide in areas with heavy pedestrian traffic [14], while pavements around shops or bus stops should be at least 3,000–3,500 mm [15]. Odufuwa [2] found that pavements in Nigeria were in a deplorable state as the pavements were not paved and also crowded with street vendors. Pavements which are characterised by crowded street vendors are common in African countries, for example, Nigeria [2], Ghana [11] and Zimbabwe [12].

Street crossings are form part of road and pedestrian environment which pose danger to people with disability, especially people with mobility, visual and hearing disability [16]–[18]. According to the best practices, street crossing should have a minimum with of 1,200 mm and centre islands are recommended in areas where pedestrian cross many lanes of traffic [14]. In Nigeria, Bombom and Abdullahi [16] found that busy streets are difficult to cross for people with disability, especially those travelling without travel assistants.

Street furniture is defined as "all the setups that are found on the streets that serve many functions related to the usage of the streets in many ways whether directly or indirectly, and that leads to better usage of the streets as an urban space been used by people" [19]. Street furniture includes benches, bins, bus shelters, signboards, water fountains and other elements that make pedestrians comfortable [19], [20]. One of the important factors to consider around street furniture is their placement on the street [21]. Street furniture should not obstruct the flow of pedestrian and objects such as benches should be provided at regular interval of every 50 m [14], [20]. Research done by Lid and Solvang [22] in Norway, reveals that in areas which had benches on the street, people in wheelchairs found it difficult to manoeuvre. Bus stops are vital elements to the pedestrian environment. According to the best practices, bus stops should be close to amenities and transport users should not walk more than 400 m along a route [14], [15]. Bus stop shelters should have sufficient space for wheelchair to enter and manoeuvre [14]. In Stockholm, transport facilities such as bus stops are designed with universal access features and in urban areas bus stops have elevated kerbs [23]. However, in Chengdu, China, it was observed that the bus stops do not accommodate the needs of people with visual disability [17]. Bjerkan et al. [24] found that one of the main challenge of passengers who use public transport was getting to the bus stop.

Bus and train stations should be designed with accessible features to accommodate different groups transport users, for example, design of entrances, walkways, ticket counters, information, amenities and platforms [14], [15], [25]. According to the best practices, the layout of stations should be simple and compact to reduce walking distances and making navigation easy for everyone [14], [25]. Ideally, all entrances of stations should be fully accessible [14], [25]. Metrorail in South Africa is not designed for people with disability [26]. Similarly, in Pakistan, stations are not universally designed to cater for the needs for all [27]. In Stockholm, Stjernborg [23] found that some lifts at the stations can be out of order for two or three months, thereby making it impossible for people using wheelchairs and others with mobility disabilities to reach platforms.

Taxi ranks in this study, refers to the transport facilities for minibus taxis commonly found in African countries. The mini-bus taxi industry plays a vital role in public transport in South Africa. According to the National Households Travel Survey 2020, about 66% of households in South Africa use mini-bus taxis [28]. The two types of taxi ranks are on-street and off-street taxi ranks. On-street taxi rank can be described as "a place usually within the road reserve at which minibus-taxis are allowed to stop for passengers to board and alight the vehicle" [29], while off street taxi ranks are usually found in a building. Ideally, off-street taxi ranks should have pavements, shelters, waiting areas and amenities such as toilets and shopping facilities [29]. One of the factors considered concerning on-street taxi ranks is width of pavement. Pavements should have sufficient width to avoid obstruction to pedestrians by waiting taxi passengers, especially when located on a busy street.

3 RESEARCH METHOD

This study was conducted in CBD of Tshwane. The City of Tshwane is of the largest metropolitan municipalities in Gauteng province of South Africa. The City of Tshwane has a population of approximately 3,600,000 and area coverage of 6,345 km^2 [30]. Because of spatial planning of the city, many residents of Tshwane spend long time commuting between their homes and places of employment [30]. "The City of Tshwane still resembles the apartheid city form, with most black townships on the outer-skirts of the city, namely, Mamelodi to the east, Atteridgeville to west and Hammanskraal and Soshanguve to the south" [30]. Feature characteristics of Tshwane CBD include main transport terminals, transport arteries, high activity of street vendors, lack of space and the presence of many national government departments [9]. The main streets for public transport in which data were collected were Madiba Street (Vermeulen Street), Pretorius Street; Lilian Ngoyi Street (Van der Walt Street); and Sisulu Street (Prinsloo Street). Data for taxi ranks and bus stations were collected from Marabastad rank.

To address research objectives, qualitative research was adopted, using semi-structured observation. Observation is one of the research techniques widely used for access audit [31], [32]. An observational checklist was prepared to collect data, and the checklist was based on guidelines and best practices for universally accessible transport [14], [15], [20], [25], [33], [34]. The guidelines of accessible design for infrastructure specify the desirable requirements or minimum dimensions of infrastructure elements, for example, width of doors, slope of gradient of ramps, platforms, width of pavement or colour. The infrastructure elements considered in this study are pavements, street crossings, street furniture, bus stops, bus station, taxi ranks and train stations. To capture pictures and videos of the condition of infrastructure around Tshwane CBD, a mobile phone was used. Field notes concerning the condition of infrastructure were also written.

4 RESULTS

The results are divided into seven themes, namely pavements, street crossings, street furniture, bus stops, taxi ranks, bus stations and train stations.

4.1 Pavements

The variables which were observed on the pavements around the Tshwane CBD are availability, width, type of surface, maintenance, obstructions, resting areas, tactile surfaces, signage, dedicated bicycle lanes, continuity of paths, buffer space, cleanliness, provision of amenities, and speed management. Fig. 1 shows obstructions on pavements around the CBD.

Figure 1: Pavement obstructions.

It was observed that mini-bus taxis, vendors, shop products and bins were obstructing free flow of traffic. The width of some pavements was reduced by street vendors, minibus taxis and shop products which would make it difficult for people with vision and mobility disability to manoeuvre.

Fig. 2 shows broken pavements and incomplete construction of pavements. From the observations it was found that some broken pavements have been in that state for a long time, while some newly constructed pavements were incomplete. Fig. 3 shows the state of tactile surfaces in some areas around the CBD of Tshwane. The findings show that the tactile surfaces around the CBD were mostly in poor condition and lack maintenance. To a large extent, tactile surfaces are not functionally existent around the CBD.

It was observed that some pavements are not kept clean as shown in Fig. 4. The findings also revealed poor spacing of bins around Tshwane CBD. Fig. 5 shows the state of some of the signage around Tshwane CBD. The signage board in Fig. 5 is damaged. From observation, the damaged signage board was in that state for a long time which shows lack of maintenance of infrastructure. The findings also revealed that some signboards are obstructed by street light poles and traffic light poles.

Figure 2: Maintenance of pavements.

Figure 3: Tactile surface.

4.2 Street crossings and street furniture

People with disability often encounter problems when crossing streets. According to the best practices, traffic signal should have push buttons within reach, audible and visual signals and countdown timers [14]. Street crossing should have a minimum with of 1,200 mm and centre islands are recommended in areas where pedestrian cross many lanes of traffic [14]. From observation, the traffic light appeared to be in that state for a long time and it was common

Figure 4: Cleanliness of pavements.

Figure 5: Signage.

around the CBD. The traffic lights around the CBD of Tshwane did not have countdown signals and a few had push buttons that were not functional.

Street furniture includes bins, water fountains, benches and street lights. According to best practices, street furniture should be correctly positioned with a contrasting colour to the surroundings [14]. Fig. 6 shows the state of some of street lights around Tshwane CBD. The light is broken and this was common around the CBD. It was found that street furniture lacked maintenance. Some of street furniture such as lights and bins were poorly spaced and positioned. It was also observed that some street lights are left on during the day. Some bin containers were full of rubbish and the containers looked dirty. Around the CBD of Tshwane, only a few seating areas were observed.

Figure 6: Traffic light and street light.

4.3 Bus stops

Bus stops should be accessible to everyone with transport information displayed inside and outside the shelter. Bus shelters should have seats and sufficient space for wheelchairs to enter [14]. The findings showed that bus stops are the CBD do not provide any transport information. Only a few bus shelters were identified around the Tshwane CBD

4.4 Bus station

The two types of bus stations audit were the Bus Rapid Transit (BRT), A Re Yeng, bus stations and the stations for conventional buses which are located on the perimeters of the Tshwane CBD. Fig. 7 shows one of the A Re Yeng bus station in the CBD. The findings showed that, to a large extent, the A Re Yeng bus stations were accessible and designed to accommodate different types of transport users, including people with disability. It was observed that starting from the entrance to the exit, the design was fully accessible to people with disability and those with reduced mobility. Although the stations were fully accessible, the outside surrounding environment of some of the stations posed barriers to people who use wheelchairs. It was observed that some pavements leading to the stations did not comply with universally designed principles.

The findings showed significant differences in the level of access of Marabastad bus stations and A Re Yeng bus stations. During the time the observation audit was conducted, the City had done some maintenance work on Marabastad stations (Fig. 8). It was observed that station shelters, pavements and handrails were in good condition. The flow of the station layout was simple and easy to follow. However, it was observed that the station platforms

Figure 7: A Re Yeng bus station.

Figure 8: Marabastad bus station.

did not have ramps and some buses were loading in the middle of a road. Luggage was obstructing the free flow of passengers. There were no ticket counters at Marabastad bus station that were identified.

4.5 Mini-bus taxi ranks

The two types of taxi ranks for mini-buses around Tshwane CBD are off-street and on-street taxi ranks. It was observed that some of the off-street taxi ranks did not have pavements and other amenities such a s toilets and shopping facilities. Off-street taxi ranks had uneven gravel surfaces which was not ideal for people using wheelchairs. One of the basement off-street taxi ranks did not good lighting and ventilation. Fig. 9 shows a taxi loading zone.

Figure 9: Taxi loading zone.

It was observed that street vendors obstruct taxi entrances which was not ideal for people using wheelchairs and people who have visual disability. The vendors' tables were next to the taxi entrance as shown in Fig. 9. The findings also revealed a lack of transport information and signage.

4.6 Train stations

The main types of train stations for passengers in the City of Tshwane are Metrorail stations and Gautrain stations. Metrorail is a division of a state-owned enterprise, Passenger Rail Agency of South Africa (PRASA), which operates commuter rail services in eight metropolitan areas in South Africa [35]. Gautrain is a rapid rail that connects Tshwane, Johannesburg and OR Tambo International Airport in the Gauteng province of South Africa. The study only focused on Metrorail stations as Gautrain stations were newly designed with universal access features. It was observed that some station entrances were fully accessible while some were not accessible. The design of stations lacked consistency. Some stations were step-free while some were not. The floor of some stations was damaged and lacked maintenance. Some handrails outside the stations were broken and posed dangers to everyone. The older rolling stock of Metrorail known as "yellow trains" were found to be non-complaint with universal access principles. The trains floor and platform were not at the same level. There were no tactile surfaces in the station which could endanger people with vision disability.

5 DISCUSSION

The discussion on findings is divided into seven variables namely: pavements, street crossings, street furniture, bus stops, taxi ranks, bus stations and train stations.

5.1 Pavements

The pavements around the Tshwane CBD were of sufficient width and complied with minimum width requirements. However, some pavements were mainly obstructed by street vendors, vehicles parked on pavements, rubbish bins and shop products. Obstructed pavements make mobility and manoeuvring difficult especially for people with vision and mobility disability. The findings are similar to other African countries such as Ghana, Nigeria and Zimbabwe [2], [11], [12], [36]. The pavements generally lacked tactile surfaces which makes mobility of people with vision disability difficulty and constantly depend on travel assistants. Although the surface finish of the pavement was found to be of acceptable quality, some were broken and showed lack of maintenance. In some areas some pavements were not levelled and some manhole covers were missing. The pavements were fairly clean; however, litter was spotted in some areas which could suggest poor spacing of bins. Previous research indicates that uneven surfaces make movement difficult for people with vison disability [3] and people who use wheelchairs [22].

5.2 Street crossing

Overall, the assessment of street crossings around the Tshwane CBD suggests that street crossings are not safe and are relatively inaccessible for people with disability. This finding is in line with research done in Nigeria which reveals that some streets which are difficult to cross for people with disability, especially those without travel assists [16]. The street crossings around Tshwane CBD were not designed for people with disability. The traffic lights around the CBD do not have countdown signals and the few push buttons which were spotted were not functional. The state of traffic lights around the CBD have adverse effect on people with mobility, hearing and vision disability. According to the best practices on street crossings, traffic signals should be accessible to everyone and should have audible and visual signals and countdown timers [14], [20]. There is a high risk for people with disability to collide with fast moving vehicles [16].

5.3 Street furniture

According to Radwan and Morsy [19], variables such as "line, texture, colour and measurement" should be taken into consideration when planning and designing street furniture. Generally, there is limited street furniture around Tshwane CBD and the existing furniture lacks maintenance. Some street furniture was poorly placed and spaced, for example, bin which were in the middle of the pavement and street light poles obstructing signage. Similarly, Lid and Solvang [22], found that benches around seating areas were not properly placed and wheelchairs users found it difficult to manoeuvre around areas where benches were available.

5.4 Bus stops

The design of bus stops around the CBD did not comply with universal design principles. Similarly, in Pakistan, Ahmad [27] indicates that existing bus stops are not universally accessible. Bus stops did not have transport information and timetables were relatively nonexistent. Coxon et al. [18] highlight that availability of bus timetables and routes forms part of small changes which can make significant impact to transport users. Research done in Stockholm reveals that although bus timetables are available, the format of presenting information was not accessible to everyone [23]. There were only few bus shelters around the Tshwane CBD of which, were not designed to accommodate people who use wheelchairs. Some bus stops did not have clear markings showing no-parking zones. Overall, bus stops were fairly clean and bins were placed around the bus stops.

5.5 Bus stations

The two types of bus stations which were audited were for conventional buses and BRT. As expected, the BRT station were found to comply universal design principles while conventional bus stations were not. A Re Yeng buses stations have various feature such information displays, ramps, waiting areas and information announcements. However, the surrounding environment leading to the BRT stations did not comply with universal access principles. Some surfaces had gravel and appeared to be uneven which makes mobility of people who use wheelchairs difficulty. The stations for conventional bus lacked ticket counters of which, is an inconvenience to transport users. However, at time of data collection, the stations for conventional buses had been renovated and showed some improvements in certain areas, for examples, legible signage, handrails and clean seating areas.

5.6 Taxi ranks

The overall assessment on taxi ranks revealed the design of taxi ranks does not comply with universal design principles. Although in South Africa, the National Building Regulations and Building Standards Act (103 of 1977) stipulate that transport infrastructure such as taxi ranks should comply with the building regulations, the design of many taxi ranks is non-compliant. This finding is in line with previous research done in South Africa that reveals that existing transport infrastructure such as taxi ranks are not universally accessible [37]. Some of the off-street taxi ranks had uneven surfaces made of gravel. Basement taxi ranks did not have sufficient lighting and security. The loading zones for taxis were non-compliant with universal design principles. On-street taxi ranks were generally located along busy streets and the taxis obstruct free flow of traffic.

5.7 Train stations

The design of Metrorail stations does not comply with universal design principles. This finding is in line with the research done in Cape Town indicating how inaccessible the Metrorail station were to people who use wheelchairs [26]. The rolling stock which was operating during the time data were collected had floors which were not at the same level with platforms. The platform-to-vehicle gap is one of critical barriers encountered by people with mobility and vision disability [38], [39]. The surfaces around the Metrorail stations and the platforms did not have tactile warning indicators or surfaces which poses high risk of falling of people with vision disability.

6 CONCLUSIONS

The overall assessment of the Tshwane CBD revealed that the design of transport infrastructure does not comply with universal design principles and, in areas where there was compliance, there was a distinct lack of consistency. The urban transport infrastructure around the CBD is not designed to cater to the needs of people with disability. The findings also revealed that existing transport infrastructure lacks maintenance. The findings suggest that the city does not support independent mobility of people with visual disability as the lack of consistency in the implementation of the tactile surface was consistent throughout the CBD. The study also found that in busy areas, pavements were overcrowded with street vendors, taxis parked on the pavement and shop owners displaying their products on the pavement. However, pavement obstruction does not only affect people with disability but everyone. The city is recommended to train city planners and built environment professionals on disability building regulations. City planners and built environment professionals should be conscious of the needs of different groups of people when planning and designing city infrastructure. The City of Tshwane it is recommended to involve people with disability through participation in planning, designing and implementation of transport infrastructure, transport policies and legislation.

ACKNOWLEDGEMENTS

Special thanks to Professor R. Luke, my PhD supervisor, Gwinyai Duri and the Department of Transport and Supply Chain Management of the University of Johannesburg for making completion of this research possible.

REFERENCES

[1] Hernández, D., Public transport, well-being and inequality: Coverage and affordability in the city of Montevideo. *CEPAL Review*, **122**, pp. 151–172, 2017.

[2] Odufuwa, O.B., Towards sustainable public transport for disabled people in Nigerian cities. *Studies on Home and Community Science*, **1**(2), pp. 93–101, 2007.

[3] Park, J. & Chowdhury, S., Investigating the barriers in a typical journey by public transport users with disabilities. *Journal of Transport and Health,* **10**, pp. 361–368, 2018. DOI: 10.1016/j.jth.2018.05.008.

[4] Cepeda, E.P., Galilea, P. & Raveau, S., How much do we value improvements on the accessibility to public transport for people with reduced mobility or disability? *Research in Transportation Economics*, **69**, pp. 445–452, 2018.

[5] Chiwandire, D. & Vincent, L., Wheelchair users, access and exclusion in South Africa higher education. *African Journal of Disability*, **6**(0), a353, 2017.

[6] Thomas, D.P., Public transportation in South Africa: Challenges and opportunities. *World Journal of Social Science Research*, **3**(3), pp. 352–366, 2016.

[7] Statistics South Africa (Stats SA), Community survey 2016. Provincial Profile Gauteng. www.statssa.gov.za. Accessed on: 30Jan 2019.

[8] Grobbelaar-du Plessis, I. & Grobler, C., *South Africa. African Disability Rights Yearbook*, pp. 307–340, 2013. https://home.heinonline.org/. Accessed on: 7 Feb. 2019.

[9] Boshoff, D.G. & Marais, D., The effects of the expanding outskirts of Tshwane on its CBD and internal structure. *Proceedings in Advanced Research in Scientific Areas* (ARSA), **2**(1), pp. 13–26, 2013.

[10] Wojnowska-Hecial, M., Heciak, J. & Klak, A., Concrete paving slabs for comfort of movement of mobility-impaired pedestrians: A survey. *International Journal of Environmental Research and Public Health*, **19**, p. 3183, 2022.

[11] Owusu-Ansah, J.K., Baisie, A. & Oduro-Ofori, E., The mobility impaired and the built environment in Kumasi: Structural obstacles and individual experiences. *Geography Journal*, **84**, pp. 1003–1020, 2019. DOI: 10.1007/s10708-018-9907-y.

[12] Ndebele, T., Neglect and discrimination: A tale of Zimbabwe inaccessible public transport system. *Journal of African Studies and Development*, **12**(3), pp. 97–103, 2020. DOI: 10.5897/JASD2020.0579.

[13] Vanderschuren, M.J., Baufeldt, J. & Phayane, S., Mobility barriers for older persons and people with universal design needs in South Africa. *14th International Conference on Mobility and Transport for Elderly and Disabled Persons*, 2015.

[14] United Nations Development Programme (UNDP), A review of international best practice in accessible public transportation for persons with disabilities. UNDP: Malaysia, 2010.

[15] Sze, N.N. & Christensen, K.M., Access to urban transportation system for individuals with disabilities. *IATSS Research*, **41**, pp. 66–73, 2017.

[16] Bombom, L.S. & Abdullahi, I., Travel patterns and challenges of physically disabled persons in Nigeria. *Geography Journal,* **81**, pp. 519–533, 2016. DOI: 10.1007/s107 08-015-9629-3.

[17] Wu, C., Li, J. & Li, X., Using community planning method to improve effect of urban barrier-free transportation system. *Transportation Research Procedia*, **25**, pp. 4330–4337, 2017.

[18] Coxon, S., Napper, R. & Richardson, M., *Urban Mobility Design*, Elsevier, 2019.

[19] Radwan, A.H. & Morsy, A.A.G., The importance of integrating street furniture in the visual image of the city. *International Journal of Modern Engineering Research*, **9**(2), pp. 29–53, 2016.

[20] Australian Government, Accessibility design guide: Universal design principles for Australia's aid program. A companion volume to development for all: Towards a disability-inclusive Australian aid program 2009–2014. www.ausaid.gov.au/publications. Accessed on: 28 Mar. 2020.

[21] Davis, C.J., *Street Design for All: An Update of National Advice and Good Practice*, Public Realm Information and Advice Network, 2014.

[22] Lid, I.M. & Solvang, P.K., (Dis)ability and the experience of accessibility in the urban environment ALTER. *European Journal of Disability Research*, **10**, pp. 181–194, 2016.

[23] Stjernborg, V., Accessibility for all in public transport and the overlooked (social) dimension: A case study of Stockholm. *Sustainability*, **11**(18), p. 4902, 2019. DOI: 10.3390/su11184902.

[24] Bjerkan, K.Y., Nordtømme, M.E. & Kummeneje, A., Transport to employment. *Scandinavian Journal of Disability Research*, **15**(4), pp, 342–360, 2013.

[25] Ahmad, M., Independent-mobility rights and the state of public transport accessibility for disabled people: Evidence from southern Punjab in Pakistan. *Administration and Society*, **2**, pp. 197–213, 2015. DOI: 10.1177/0095399713490691.

[26] Coxon, S., Napper, R. & Richardson, M., *Urban Mobility Design*, Elsevier, 2019.

[27] Ahmad, M., Independent-mobility rights and the state of public transport accessibility for disabled people: Evidence from southern Punjab in Pakistan. *Administration and Society*, **2**, pp. 197–213, 2015. DOI: 10.1177/0095399713490691.

[28] Statistics South Africa (Stats SA). National households travel survey 2020. Statistical release P0320. Accessed on: 14 May 2021.

[29] Wagner, L.P., The siting of public transport facilities in a predominately low to middle-income area, Klapmuts, on the outskirts of a metropolitan city, Cape Town. Masters dissertation, University of Cape Town, 2017.

[30] Cooperative Governance and Traditional Affairs, City of Tshwane Metropolitan GAU. https://www.cogta.gov.za/ddm/wp-content/uploads/2020/11/Tshwane-October-2020.pdf.

[31] O'Hanlon, J., Scott, M.S. & West, L., Healthy and complete communities in Delaware: The walkability assessment tool. Institute for Public Administration, University of Delaware, 2016.

[32] Galanis, A. & Eliou, N., Development and implementation of an audit tool for the pedestrian built environment. *Procedia – Social and Behavioral Sciences*, **48**, pp. 3143–3152, 2012.

[33] European Commission, Best Practice Guide. 20 Best Practice Examples Promoting Sustainable Urban Mobility. https://mobilityweek.eu. Accessed on: 13 Jan. 2020.

[34] European Conference of Ministers of Transport (ECMT), Improving transport accessibility for all: Guide to good practice. OECD Publications Service: France. 2006.

[35] Stjernborg, V., Accessibility for all in public transport and the overlooked (social) dimension: A case study of Stockholm. *Sustainability*, **11**(18), p. 4902, 2019.

[36] Ipingbemi, O., Mobility challenges and transport safety of people with disabilities (PWD) in Ibadan, Nigeria. *African Journal for the Psychological Study of Social Issues*, **18**(3), pp. 15–28, 2015.

[37] Lister, H.E. & Dhunpath. R., The taxi industry and transportation for people with disabilities: Implications for universal access in a metropolitan municipality. *Transformation*, **90**, 2016. DOI: 10.1353/trn.2016.0009.

[38] Oksenholt, V.K. & Aarhaug, J., Public transport and people with impairments: Exploring non-use of public transport through the case of Oslo, Norway. *Disability and Society*, **33**(8), pp. 1280–1302, 2018. DOI: 10.1080/09687599.2018.1481015.

[39] Zajac, A.P., City accessible for everyone: Improving accessibility of public transport using the universal design concept. *Transportation Research Procedia*, **14**, pp. 1270–1276, 2016.

EVALUATING THE ACCESSIBILITY OF A SUSTAINABLE CITY: CASE STUDY OF MASDAR CITY, UAE

PRAVEEN KUMAR MAGHELAL[1], XIANGYU LI[2] & ANOUD QAHTANI[3]
Radban Academy, UAE
University of Oklahoma, USA
Khalifa University, UAE

ABSTRACT

An epitome of a sustainable city, Masdar City in the United Arab Emirates, boasts of its integrated land use and multi-modal transportation network. When developed as planned, it could be an example for designing future cities. However, little is known of how accessible it could be in spite of its unique design. Conversely, accessibility, as a conceptual and operational definition, has gone through several iterations. However, it still remains an often misunderstood and poorly defined and measured construct. Masdar City, as planned, consists of several types of sustainable transportation options including walking, biking, personal rapid transit (PRT), group rapid transit (GRT), bus/light rapid transit, and metro. The hierarchical additive approach (considering both spatial analysis and the time-based analysis), reports that majority of Masdar City has low accessibility to the regional metro station. Also, the result indicates that more than half of the city will need between 13.1 to 25 minutes, with an average of 15 minutes, to reach the metro station. In all, the city has medium accessibility to the metro station with opportunities to enhance the overall accessibility. Elimination of transit stops that are redundant can reduce the headway and stalling time and hence the total travel time. Also, by locating new transit stops (bus, PRT or LRT) in grids where access to closest public transit is more than 400 m by walking can help balance the capital cost of infrastructure (public transit) while improving the accessibility to the regional metro for its residents. The follow-up analysis reported that the medium accessibility increased from 20% to 24%. Evaluation of accessibility in planned areas (city or neighbourhood) can provide insights on efficiency of a sustainable city such as Masdar City.
Keywords: sustainable city, Masdar City, accessibility, United Arab Emirates, transportation network.

1 INTRODUCTION

Major cities around the world are facing unprecedented challenges that might change its demographic, environmental, economic and social aspects owing to an exponential increase in urban population [1]. It is projected that in the next 15 years the world population would continue to grow by more than one billion a year to about 8.5 billion in 2030; 9.7 billion in 2050; and about 11.2 billion by 2100 [2]. Most of this growth is projected to happen in urban areas and its periphery, which has followed the trend of sprawling over the last six decades.

Urban sprawl, term use to describe the complex undesired urban expansion pattern, is a phenomenon that is a result of the rapid population growth (both total and urban). These expansion areas are poorly designed neighbourhood forcing its dwellers to use cars for most of their trips – including the short trip to the grocery – that result in higher automobile dependency [3]. Urban designer, planners and architects have proposed several strategies and approaches to combat this ad hoc developmental approach around the world.

Planning paradigm such as smart growth, new urbanism and transit-oriented development are some approaches proposed to design and develop sustainable cities. These approach encourages diversity in land-use, compact development, diversity in housing type, walkable neighbourhoods, create a sense of community, preserve open space, farmland, and critical habitat, infill development and offer a variety of transportation modes [4].

Cities have been assessed for its shortcomings and success on adapting these approaches at various scales from regional, local to neighbourhood level and for its implication on land-

www.witpress.com, ISSN 1743-3509 (on-line)
doi:10.2495/UMT220081

use planning, transportation services, climate change, social capital, and health outcomes [5]–[7]. However, little research exists that analyse a city planned to be developed for its implications on these outcomes. Such investigations can provide insights to recommend changes that can enhance the social, environmental and economic liveability of future habitats of cities around the world.

Masdar City, in the Abu Dhabi Emirate of the United Arab Emirates is proposed as a carbon neutral city and has been identified as an epitome of sustainable city. The city, designed by renowned architect and planner Norman Foster, planned it to be a sustainable development benchmark and a green-print for how cities can manage the rapid urbanization and reduce the consumption of energy, water, and waste impact on the environment with the renewal source of energy [8]–[10]. This city, as planned, boasts of mixed-use development which will include residential, retail, educational, recreational, light industrial, and office areas, which will provide a place to live–work–play for its residents. Moreover, the city is designed to offer a friendly pedestrians and cycling environment with an integrated smart transportation network that makes the city more accessible and liveable. This integrated smart transportation network hopes to provide efficient accessibility for its users across various destinations within the city [11]. However, while high accessibility can be presumed for a new development such as Masdar City, its assessment can provide insights into the efficiency of this integration of transportation network.

This study therefore uses measure of accessibility using the transportation components to evaluate the percentage of area most to least accessible through various modes of travel proposed for the city. Furthermore, we discuss implication and recommendation to enhance the accessibility of the city. Implications target the reduction in lifecycle cost of transportation infrastructure that can reduce fiscal impact of planned and unplanned cities in future.

2 MEASURES OF ACCESSIBILITY

Scholars define accessibility as "the potential of opportunities for interaction" [12]. It is further defined as the ability and opportunities of reaching desired products, services, activities or destinations [13]. Unlike mobility that focuses on movement, accessibility focuses on making products and goods closer to the consumers, such as improving the accessibility to schools and retail stores by locating them in the middle of a neighbourhood and accommodate the needs of people by reducing the travel distance and time [13].

Therefore, accessibility measures help to predict the impacts of any transportation investment and evaluate their effectiveness on the dweller's daily behaviour. It provides insights to planners and decision makers to assess the distribution of activities, land use and transportation facilities and services [14], [15].

While several measures of accessibility have been proposed [16], Yigitcanlar et al. [17] developed a GIS-based land use and public transport accessibility indexing model (LUPTAI). LUPTAI is a comprehensive – origin-based accessibility-indexing model that targets to measure the ease of accessing common land use destinations such as residential, commercial, health and retail either by walking or using public transit. It is one of the first models that consider public transit as the methods of access, not a service to be accessed. It relies on information about land-use destinations, road network, pedestrian routes and public transit network. The main aim of this model is to assist in evaluating the accessibility to specific destinations within the city planning and the decision-making phases.

3 METHODS

The accessibility measure developed by Yigitcanlar et al. [17] and modified by Albacete et al. [12] was identified as the most appropriate method to assess accessibility with Masdar City for two reasons. Firstly, Masdar City, as planned, is yet to be developed to the extent to which we could assess the temporal and individual components of accessibility [16]. Without any residents, it would be impossible to assess these components. Secondly, the development as proposed will occur in phases while the transportation network remains as planned irrespective of any change in land-use. Hence the accessibility measure using the location-based approach was applicable for a city at its infancy of development.

Table 1 indicates the range and approach of reporting the level of accessibility for each grid. Using the same standards and measures of distance and time, access to the only metro station within the city that connects to the mainland Abu Dhabi (major source of employment and hence possible destination for intercity travellers) was evaluated and mapped in GIS. The procedure for the same is described in the following section.

Table 1: Walking, public transit, and land use accessibility measure scores. *(Source: Adapted from [12], [17].)*

Layer	High 4	Medium 3	Low 2	Poor 1	Very poor 0
Walking to transit (in meters)	0–300	301–400	401–800	801–1200	> 1201
Total time to public transit (in minutes)	0–6	6.1–13	13.1–25	25.1–38	>38.1

3.1 Grid cells assessment

A grid of 400 m × 400 m was spatially laid over the city area in GIS as unit of accessibility measure (Fig. 1). According to the transit-oriented development (TOD) guidelines, the common standards, a 5-minute walk (equivalent to 400 m) is an acceptable walking distance/ time to access public transit with a walking speed of 5 km/h or about 80 m/min [18]. Also given the hot arid temperature in UAE, several studies use this measure as acceptable walking distance for their analysis [19], [20]. Each of these cells will be assigned an accessibility score (Table 1) based on its ease of access to public transit, and various land uses in the city [12], [17], [21]. The scores ranged from 0 (being very poor accessible) to 4 (high accessibility) for each mode of travel. Based on the time to destination (metro stop) from the centre of the grid, each cell or grid was assigned an accessibility score for its respective mode of travel such as automated personal rapid transit (PRT), light rail transit (LRT) and the bus rapid transit (BRT) systems.

3.2 Spatial assessment

The spatial assessment consists of estimating the walking distance to the nearest public transit stations for each cell. This assessment was conducted for all stations and all modes of public transit to be available in Masdar City (PRT, LRT, bus, and metro). The walking distance was measured as a straight-line distance from the centre of each cell to all the closest public transit stops. Hence, the outcome of this spatial assessment was to identify which stop/station of the selected public transit mode is at the closest proximity to each grid cell centre (Fig. 2). Based on walking speed, the time to walk from the grid centre to the closest stop was calculated and categorized by its level of accessibility (by distance measure).

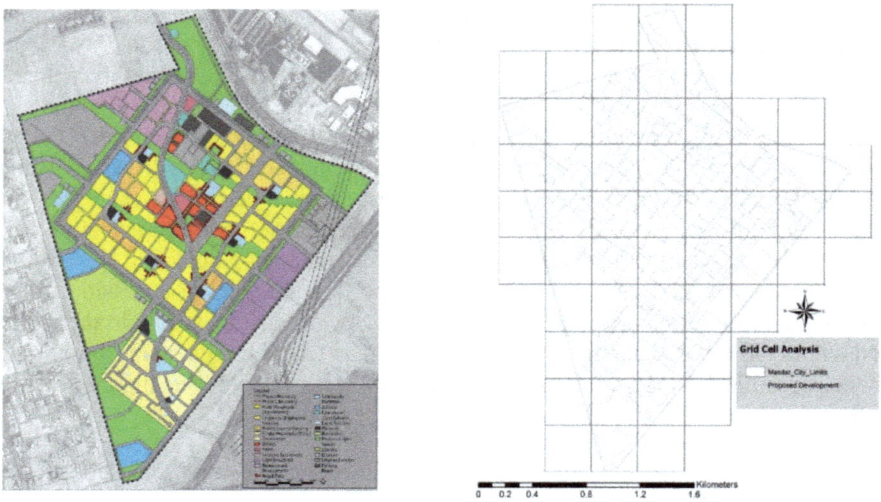

(a) (b)

Figure 1: Masdar City. (a) Master plan; and (b) Grid layout over the city.

Figure 2: Spatial assessment of accessibility to the closest transit stop: PRT, bus and LRT.

3.3 Public transit assessment: Additive approach

For the public transit analysis, the metro station in the city was selected as the public transit destination. The accessibility measured was obtained by considering the time taken to walk from the centre of the grid to the closest station or stop of any of the three public transit service, taking that transit to reach the stop closest to the metro station and final access by walk to the metro station (Fig. 3). Total transit travel time was calculated by totalling the spatial assessment (converted from distance to time), the transit travel time which included average headway time, travel time and waiting time, and walking distance from destination to metro stop (converted from distance to time). In some instances, it was observed that a transfer to another mode of transportation reported higher accessibility, in spite of transfer time and waiting time, than continuity on the same network. Based on the total time to metro station, each grid was classified using Table 1 for very poor to high accessibility.

It is apparent that grids closest to the metro station will report highest accessibility in comparison to those farther away. However, this study examines the extent of reduction in accessibility and proposes approaches to enhance the accessibility for the location farther from the metro station.

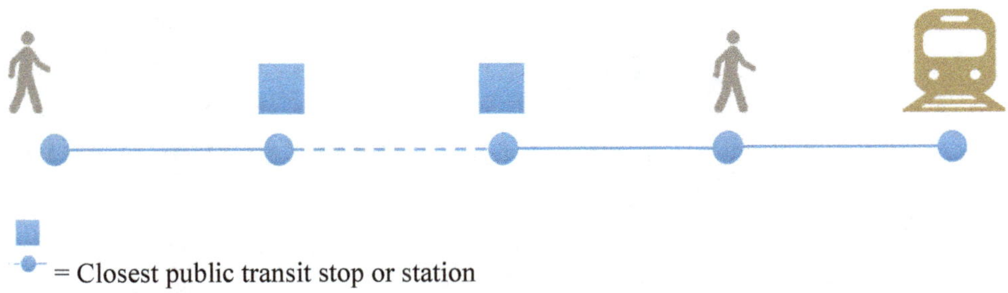

Figure 3: Additive approach to access metro station.

4 RESULTS AND DISCUSSION

4.1 Accessibility to modes of travel

The spatial assessment to estimate accessibility to each public transit was calculated, categorized, and spatially mapped (Fig. 4(a)–(c)). The PRT system report high accessibility of just 11%, while 60% of the grid cells report poor to very poor accessibility. This is justifiable given that the current network of PRT system serves only the core of the city. Without its expansion (possible at very high cost due to its automation system), the residential development (of villas) in the southern part of the city remain almost inaccessible through the PRT system.

As planned, the LRT route runs from the northwest of Masdar City to the southeast through both the large and small development squares to ensure good connectivity. While this network of LRT system reports high or medium accessibility in one-quarter of the city, the poor to very poor accessibility is around 36% of the city area. The majority of the area reports low accessibility of over 40% of the city. In comparison to the PRT system the residential development of villas in the south report better accessibility to the LRT system.

The bus network serves the majority of city and planned to be the main public transit system serving the inhabitants of the city. This network reports high to medium accessibility in about 60% of the city area. Only 6% of the city area, at the edge of the city with no planned development report poor accessibility. Most developed area, including the residential villas are well served through the bus service system.

4.2 Accessibility to metro station: Additive approach

The additive approach reported 19% of the city area with poor accessibility (Fig. 4(d)). However, upon spatial mapping of the same, all but one grid lay at the edge of the city with no proposed development. With respect to access to the metro station, about half the city reported low accessibility to the metro station and about quarter of the city reported high to medium accessibility with under 15 minutes of travel time to the metro. The recent discussion on the 15-minute city proposes such interconnected transportation and land uses, as it can enhance the liveability of cities by reducing emission and health impacts on its residents [22]. In all, this city reports decent to good accessibility, especially if the poor accessibility cells at the edge of the city with no planned development are eliminated from the overall analysis. This aligns with the vision of Masdar City to be a global example of sustainable city.

The additive approach also provided opportunity to identify stops or stations that were either redundant or unused as part of access to the metro station. Henceforth, to test the impact

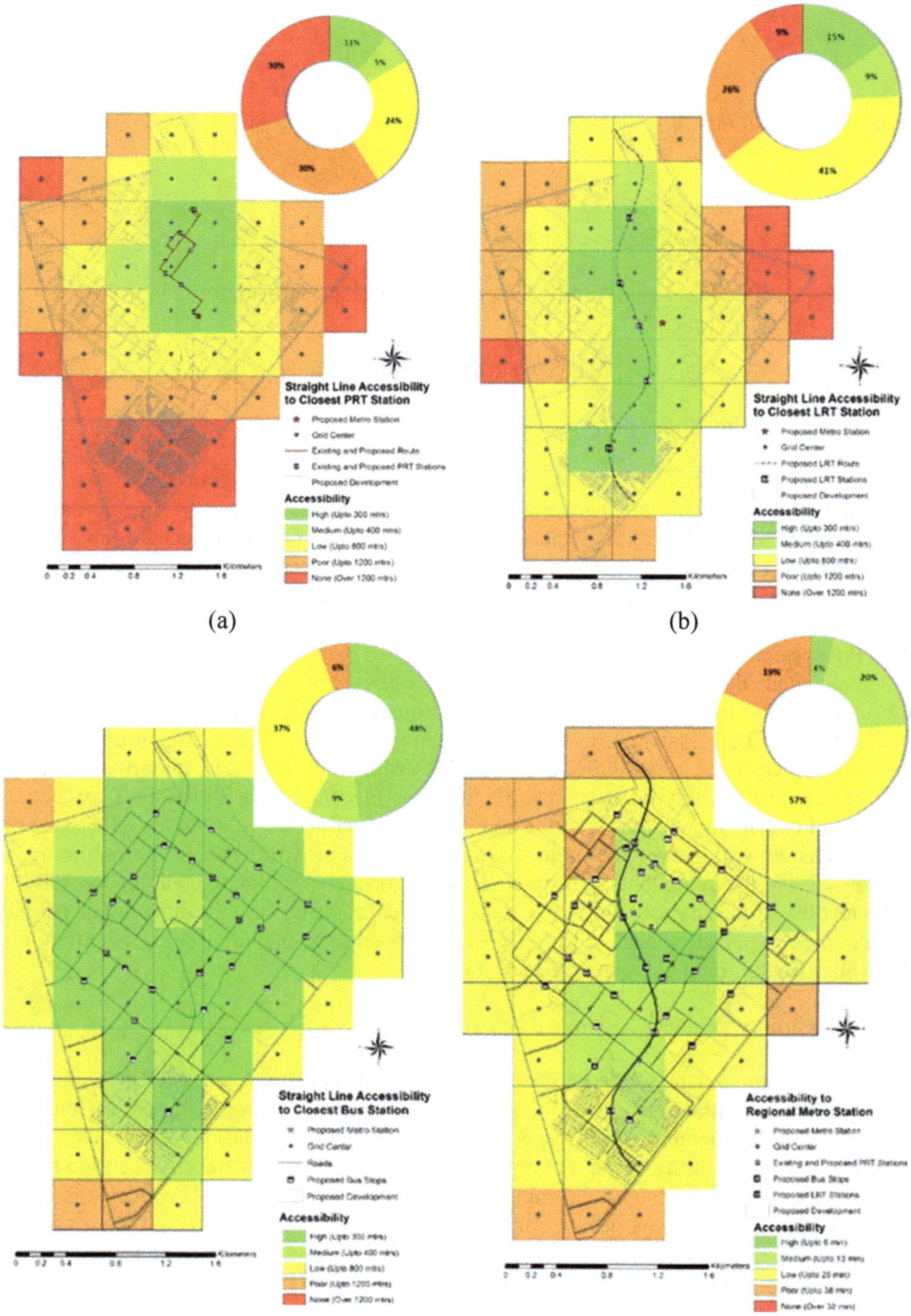

Figure 4: Measure of accessibility to metro station using: (a) PRT; (b) LRT; (c) Bus; and (d) Additive approach.

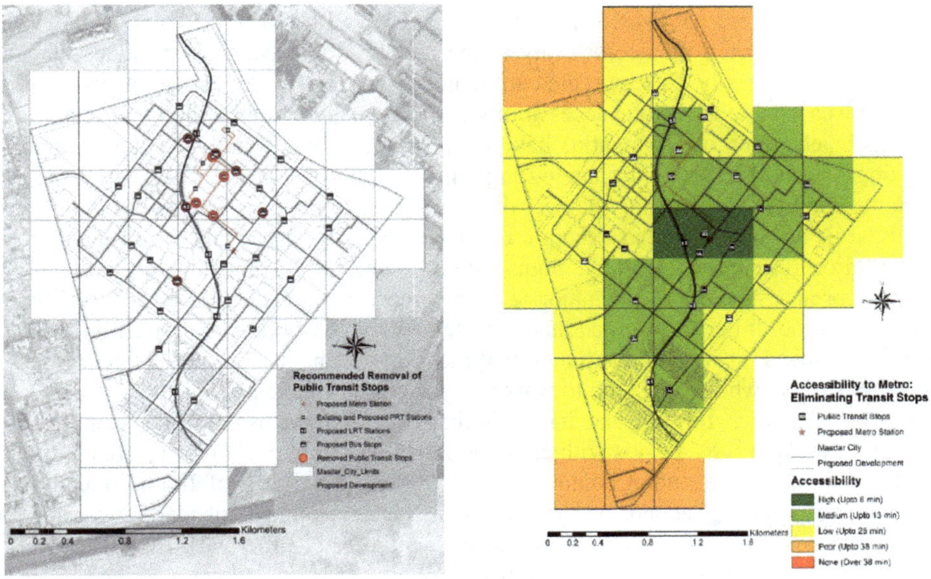

(a) (b)

Figure 5: (a) Location and type of redundant station or stop; and (b) Accessibility to metro station after removal of identified stops and stations.

of elimination of these stops on overall accessibility to the metro, five bus stops, one LRT station and four PRT stations were eliminated from the additive approach (Fig. 5(a)).

This was done to test if that improved the accessibility and if so, the benefit of elimination of these stops. The outcome reported that 17 grid cells reported improvement in accessibility (Figs 5(b) and 6). While the elimination of the LRT and PRT station did not change the level of accessibility it still reduced the time of travel to the metro station. At least five grid cells reported improvement in levels of accessibility either from poor to low or from low to medium accessibility. The total area of accessibility increase by sheer 2%. While the improvement in accessibility can be considered meagre, the elimination of these stops and station can result in tremendous saving for the city considering the life-cycle cost of development, maintenance, and operation of these transit stops, especially with negatively affecting the current accessibility of the city.

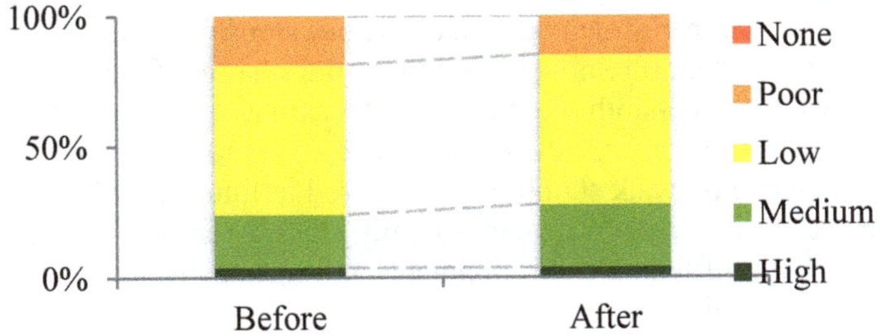

Figure 6: Change in overall accessibility in Masdar City after removal of redundant stops and stations.

5 IMPLICATIONS AND CONCLUSION

This study investigated the accessibility to the metro station using three modes of public transit service in the city. Based on the current master plan of the city, an innovative additive approach was used to assess the accessibility to the only metro station in the city, mimicing the actual behaviour of a potential habitat of Masdar City. Using existing methods of analysing accessibility and integrating those approaches, this study provides recommendations for enhancing the overall accessibility of the city. The results indicate that the city's vision to be a global example of sustainable city, especially with regards to the transportation system for this study, aligns with their objective. Almost no city area has poor or very poor accessibility, while majority of the area is within 25 minutes of travel time to the metro station. It has to be noted that, we used a very conservative approach in assessing the walking speed, as it can be assumed that most inhabitants using metro to reach the city would be workers who would tend to walk at a pace faster than 5 km/h.

As with all studies, this one also has its limitations. First, the use of grid system resulting in about 13 grids of the 54 grids with either less than half or very little of its area within the actual city limit. Future studies could use a grid system in multiple of 200 m to analyse the accessibility spatially. Second, as the city is in the development phase, the master plan presents the type of development proposed (residential or commercial). However, it is not clear if the residential development is multi-storey multi-family development and if so with what density. Finally, with the city in its infancy stage of development does not house enough residential population to evaluate the demand on transit use. Increase in population and possible demand on use of public transit could render use of some public transit service more difficult, leading to higher waiting time or skipping of the closest stop or station to reach another stop to avoid congestion.

Despite these limitations this research makes a valuable contribution by highlighting the use of accessibility measure for a planned city to inform its decision-makers about opportunities to improve future development. The specific implications from this study for infrastructure and planning a sustainable city include:

1. As a planning tool, this study emphasizes the value of assessing accessibility for newly planned communities before its development. Such an analysis can provide recommendations to improve the land uses and transit infrastructure of the new development.
2. Methodologically, this study proposed an additive method to analyse transit accessibility followed by access to various land-use destinations in the city. Although simple in concept, it is clear in sense of its implication for developing an interconnected sustainable city. As pointed out by Handy [23], a simple measure of accessibility is a way to understand the opportunities that the urban environment provides for its residents.
3. The redundant service/infrastructure such as the bus stops and PRT stations that are not easily accessible by walk, if removed can reduce the cost of infrastructure development and also improve the accessibility to various destinations within the city.
4. The land uses within the city are evolving. Hence an approach to improve the land-use mix that is accessible by walk should be considered in future developments of the city. A more comprehensive look of demand–supply side of accessibility [24] can provide implication of both sustainable development and sustainable transportation in cities targeting specific sustainable development goals.

Future research can use network distance analysis to estimate the accessibility to public transit more accurately. This study was limited with the available for such data. Also, future studies can conduct a cost–benefit analysis of eliminating the redundant stops or station for both planned and developed cities to estimate the reduction in fiscal burden for the cash-stricken cities of the world. Moving beyond the current components it is encouraged to reassess cities for its accessibility and to understand the quality of life [25] of its habitats to enhance the liveability of the cities, planned or developed. Finally, to help manage and achieve the trade-off between the quality of transit services and the transit stops accessibility, we recommend using optimization models to optimize public transit quality and system/services accessibility. Such optimization approach can help to re-allocate the stations in cities targeting the sustainable goals to achieve the trade-off and will result in maximizing the ridership coverage and minimizing travel time and cost.

ACKNOWLEDGEMENT

The authors thank the planning and transportation divisions of Masdar City for sharing the data that were used to assess accessibility within the city.

REFERENCES

[1] Hens, L., The challenge of the sustainable city. *Environment, Development and Sustainability*, **12**(6), pp. 875–876, 2010.

[2] Nations Unies, Division de la population, UN Department of Economic, United Nations, Department for Economic, Social Affairs, Population Division and Social Affairs, Population Division, *World Population Prospects: The 2000 revision*, Vol. 3, UN, 2001.

[3] Frumkin, H., Urban sprawl and public health. *Public Health Reports*, 2016.

[4] Gren, Å., Colding, J., Berghauser-Pont, M. & Marcus, L., How smart is smart growth? Examining the environmental validation behind city compaction. *Ambio*, **48**(6), pp. 580–589, 2019.

[5] Maghelal, P., Alawadi, K., Arlikatti, S. & Wahdain, A., Influence of the built environment on physical activity choices among Emirati male and female adolescents: An examination of parents' and students' perceptions. *Sustainability*, **14**(1), p. 444, 2021.

[6] Pimenta, A.R., Maghelal, P.K. & Alawadi, K., Are transit-adjacent developments effective neighborhood design models to help meet the recommended weekly physical activity levels? The case of Abu Dhabi. *International Journal of Sustainable Transportation*, **15**(3), pp. 163–174, 2021.

[7] Al-Ali, A., Maghelal, P. & Alawadi, K., Assessing neighborhood satisfaction and social capital in a multi-cultural setting of an Abu Dhabi neighborhood. *Sustainability*, **12**(8), p. 3200, 2020.

[8] Masdar City, Abu Dhabi, UAE, Masdar City solar PV plant. http://www.masdar.ae/en/energy/detail/masdar-city-solar-pv-plant. Accessed on: 22 Jun. 2022.

[9] Lau, A., Masdar City: A model of urban environmental sustainability. *Social Sciences*, **9**, pp. 77–82, 2012.

[10] Masdar City, Abu Dhabi, UAE. About Masdar. http://www.masdar.ae/en/masdar-city/live-work-play. Accessed on: 23 Jun. 2022.

[11] Transportation Master Plan. Abu Dhabi: Masdar City, 2014.

[12] Albacete, X., Olaru, D., Paül, V. & Biermann, S., Measuring the accessibility of public transport: A critical comparison between methods in Helsinki. *Applied Spatial Analysis and Policy*, **10**(2), pp. 161–188, 2017.

[13] Tumlin, J., *Sustainable Transportation Planning: Tools for Creating Vibrant, Healthy, and Resilient Communities*, Vol. 16, John Wiley & Sons, 2012.

[14] Curl, A., Nelson, J.D. & Anable, J., Does accessibility planning address what matters? A review of current practice and practitioner perspectives. *Research in Transportation Business and Management*, **2**, pp. 3–11, 2011.
[15] Handy, S.L. & Niemeier, D.A., Measuring accessibility: an exploration of issues and alternatives. *Environment and Planning A*, **29**(7), pp. 1175–1194, 1997.
[16] Geurs, K.T. & Van Wee, B., Accessibility evaluation of land-use and transport strategies: review and research directions. *Journal of Transport Geography*, **12**(2), pp. 127–140, 2004.
[17] Yigitcanlar, T., Sipe, N., Evans, R. & Pitot, M., A GIS-based land use and public transport accessibility indexing model. *Australian Planner*, **44**(3), pp. 30–37, 2007.
[18] Daniels, R. & Mulley, C., Explaining walking distance to public transport: The dominance of public transport supply. *Journal of Transport and Land Use*, **6**(2), pp. 5–20, 2013.
[19] Almardood, M.A. & Maghelal, P., Enhancing the use of transit in arid regions: Case of Abu Dhabi. *International Journal of Sustainable Transportation*, **14**(5), pp. 375–388, 2020.
[20] Alawadi, K., Hernandez Striedinger, V., Maghelal, P. & Khanal, A., Assessing walkability in hot arid regions: the case of downtown Abu Dhabi. *Urban Design International*, pp. 1–21, 2021.
[21] Papa, E. & Bertolini, L., Accessibility and transit-oriented development in European metropolitan areas. *Journal of Transport Geography*, **47**, pp. 70–83, 2015.
[22] Allam, Z., Nieuwenhuijsen, M., Chabaud, D. & Moreno, C., The 15-minute city offers a new framework for sustainability, liveability, and health. *The Lancet Planetary Health*, **6**(3), pp. e181–e183, 2022.
[23] Handy, S., Is accessibility an idea whose time has finally come? *Transportation Research Part D: Transport and Environment*, **83**, 102319, 2020.
[24] Kelobonye, K., Zhou, H., McCarney, G. & Xia, J.C., Measuring the accessibility and spatial equity of urban services under competition using the cumulative opportunities measure. *Journal of Transport Geography*, **85**, 102706, 2020.
[25] Abdullah, J. & Mazlan, M.H., Characteristics of and quality of life in a transit oriented development (TOD) of Bandar Sri Permaisuri, Kuala Lumpur. *Procedia-Social and Behavioral Sciences*, **234**, pp. 498–505, 2016.

SECTION 3
PORTS MANAGEMENT AND OPERATION

BEST PRACTICES EXCHANGE IN SEA–RAIL INTERMODALITY: A CASE STUDY OF THE PORTS OF İZMIR, TURKEY AND TRIESTE, ITALY

ZEYNEP PEHLEVAN[1] & STEFANO RICCI[2]
[1]Traffic and Station Management, Turkish State Railways, Turkey
[2]Development and Innovation in Transport Systems, Sapienza Università di Roma, Italy

ABSTRACT

Benchmarking and best practices derivation is a traditional and consolidated approach in transport planning. Less frequent is the use in the design of ports' masterplans and intermodality promotion in coastal cities, where the connection between the development of ports and local economy and the quality of the environment is stronger. Rail transport is essential for the development and efficiency of a port as it enhances economic importance and competitiveness. It also provides faster movement and higher port capacity, thus affecting the total throughput, which is a fundamental productivity measure of a seaport. This paper aims to depict a roadmap for the increase of rail modal share in ports' areas by identifying the most promising operational and infrastructural actions derived by recognized best practices. The case study described in the paper is about the exchange of best practices between the ports of Trieste and İzmir. For many years, İzmir has been an important container port in Turkey. However, its total throughput has decreased dramatically in recent years while its competitors, both in the Aegean and Marmara regions, have been developing their business rapidly. Considering that the owner and operator of the port are Turkish State Railways, the port of İzmir currently has quite a low rail share, which was less than 10% in 2021. This decrease is mainly due to an unsuitable infrastructure and poor coordination with the railway network. This paper exemplifies how it would be possible to increase the rail modal share of the port of İzmir by defining the necessary actions and redesigning the rail infrastructure. This can make the port of İzmir an effective alternative to solve the massive congestion problem of the Marmara region (Istanbul) ports, affected by morphologic barriers. Here, the port of Trieste will be the reference port as it has the highest rail share (50%) in Italy. Its current state and future targets will represent the best practice to increase the rail share in İzmir. Since almost half of the Turkish maritime freight directed to Europe is via Trieste, similar percentages on the Turkish side can strengthen the sea–rail intermodal connection between these port cities.

Keywords: sea–rail intermodality, port connectivity, railway terminal, port of Trieste, port of İzmir.

1 INTRODUCTION

Maritime transport has the highest mass carrying capacity and the lowest unit cost. This is the reason why the share of maritime transport in global trade is around 90% in volume and 70% in value. When the huge competition in seaborne trade is considered, being a *port city* is one of the most significant advantages. The *Global Cities Analysis* report published in 2019 by Oxford Economics, estimates the growth of 900 cities in the period 2020–2035. According to this, eight of the top ten cities expected to show the highest growth performance in the world were port cities [1].

In the European combined transport market, the use of intermodal load unit (ILU) shows a strong tendency toward containers, which account for about 62% of the whole market. The remaining market shares are almost equal between semi-trailers (21%) and swap bodies (17%) [2]. Thus, due to the attraction of container growing flows, it has become easier in the ports to organise efficient transport by rail only [3]. According to The European Commission White Paper, two alternatives to all-road freight transport are credible: rail transport and short-sea shipping (Fig. 1). In this context, Europe is considerably dependent on seaports.

WIT Transactions on The Built Environment, Vol 212, © 2022 WIT Press
www.witpress.com, ISSN 1743-3509 (on-line)
doi:10.2495/UMT220091

There are more than 300 active seaports in the continent; at least 130 can handle containers, of which around 40 have intercontinental container services [3].

In this paper, after getting fundamental information on maritime in Italy and Turkey, the focus is on one of the most important Italian ports to understand how to design an ideal sea–rail intermodal network. Having more than 50% rail modal share, the port of Trieste is one of the best case study in Europe. Moreover, it has also relevant maritime traffic with Turkey. After giving some more details about trade relations between those two countries, the ports of Trieste and İzmir, the last part will be clarifying what should be in İzmir to increase the rail share and be in connection with Trieste by sea–rail intermodality. Since Turkey (326 million t) and Italy (287 million t) are the top two countries in short sea shipping in Europe (Fig. 1).

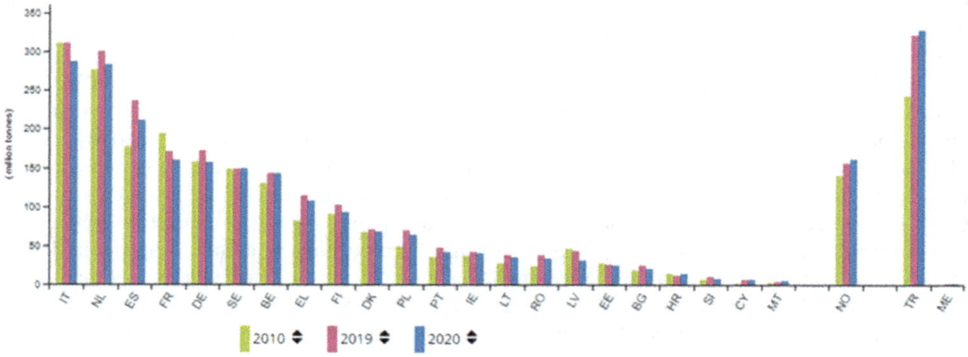

Figure 1: Short sea shipping of freight in million tons in Europe. (*Source: Eurostat.*)

2 TRADES IN MEDITERRANEAN AND POSITION OF ITALY AND TURKEY

South-east Asia and the Far East is the most crucial region regarding the number of connections [4]. Reaching the Mediterranean area from those regions, The Suez Canal plays a strategic role for global shipping as 9–10% of international trade uses it. The Mediterranean is a privileged transit route as it accounts for 27% of the 487 global liner services. This fact is also strongly affecting the ports around [5].

Turkey and Italy are important trade partners with high export and import values every year. According to 2021 data, Italy is the fourth country in export with around 11,473 million USD, while the fifth country in imports with 11,562 million USD. In the case of Italy, the trade volume between Italy and Turkey is around 17 million EUR, except for 2020, due to the pandemic restrictions.

In maritime transport between Italy and Turkey, should play an important role the port of Taranto, as one the main Turkish company, Yılport Holding, in 2020 signed a 49-year concession agreement with Port Authority of the Ionian Seaports. Taranto is the final node of the rail terminal of the Scandinavian–Mediterranean Corridor, connecting this corridor also to Valletta, Malta [6].

2.1 Overview of maritime and railway transportation in Italy

Italy is Europe's fourth largest economy and a founding member of the European Union. The country has a strong export structure and its global competitiveness, expertise superiority and innovation in significant sectors have created the *Made in Italy* brand [7]. Moreover, Italy is located in southcentral Europe, shaped as a peninsula into the Mediterranean Sea, with 7,500 km coastline [8].

Italian ports are more favourable than northern ports, especially considering navigation time from the Far East, which is about five days less than to northern range ports. The most important Italian ports (Fig. 2) can group as Ligurian ports (Genova, La Spezia, Savona, and Livorno), north Adriatic ports (Trieste, Ancona, Ravenna, and Venezia), Campanian ports (Napoli and Salerno), hub ports (Cagliari, Gioia Tauro and Taranto) and others, such as Civitavecchia, Bari, Trapani, Catania and Palermo [5]. Ligurian ports are the most convenient for Switzerland, while Adriatic ports for southern Germany and central-eastern Europe [4].

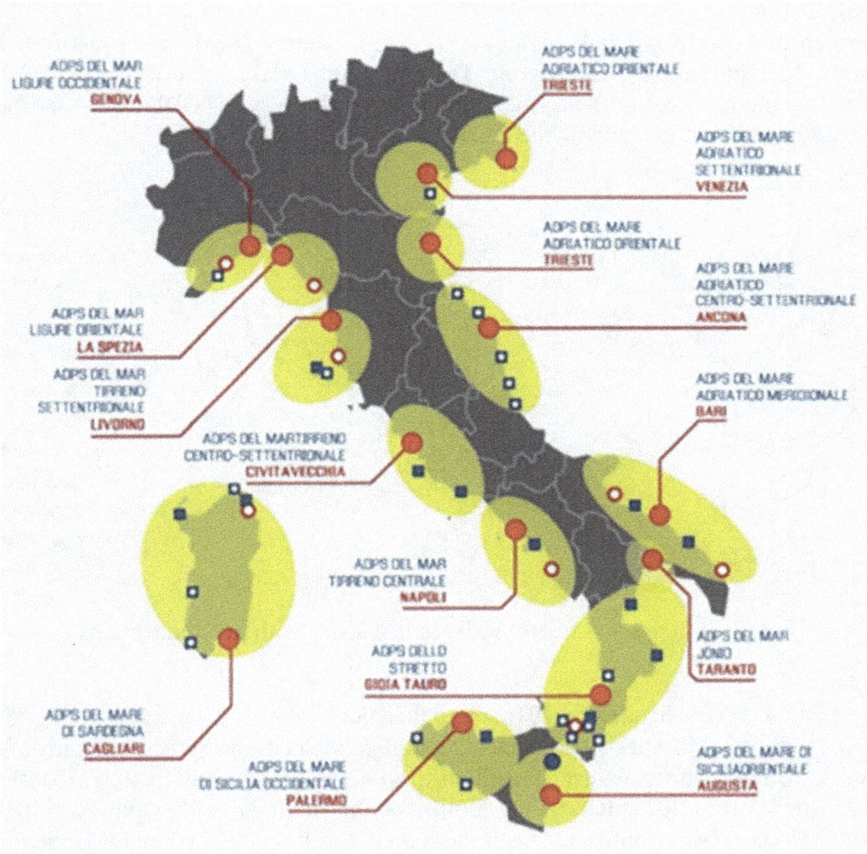

Figure 2: Italian port authorities.

There are many rail connections in operation from Italian ports, mainly from Trieste, La Spezia, Genoa and Livorno. However, still, companies are using intermodal connections marginally. The 81% of companies prefer road connection, while 19% uses a road and railway combination. The lack of rail infrastructure connected to the ports is one of the reasons for this trend [5]. Trieste registered more than 50% rail modal share; La Spezia is around 33%. Only these ports have comparable rail share with north European ports (e.g., Hamburg). The leading Italian port regarding domestic rail connections is La Spezia; meanwhile Trieste is dominating for international connections [9].

2.2 Overview of maritime and railway transportation in Turkey

Turkey is a transcontinental country between Europe and Asia. The country is geographically a peninsula with 8,333 km coastline. The maritime transport milestone in Turkey is the Treaty of Lausanne, signed in 1923 and providing the right to cabotage. In addition, Montreux Convention, signed in 1936, gave the sovereign rights of Bosporus and Dardanelles.

Maritime transport is the leading mode having around 95% share in imports and 81% share in exports in 2019. Road transport share tends to decrease in export, in parallel with the increase in maritime transport and railway, which has a share of about 1% for both import and export.

Thirteen ports have rail link in Turkey (Fig. 3), namely Haydarpaşa, Derince, İzmir, Bandırma, Mersin, Samsun, İskenderun, Tekirdağ, Zonguldak, Yılport, Evyap, DP World Ports and Nemport. Among them, the Turkish State Railways (TCDD) own and operate Haydarpaşa and İzmir, meanwhile the others became private.

Figure 3: Ports and freight villages in Turkey with rail connections.

The 5,743,455 TEU handled in 2010, more than doubled in 2021 and became 11,759,484 TEU (464,405,662 t) in Turkish ports in 2021. Again, in 2021, the highest amounts of cargo handled in Turkish ports, almost 39 million t, on 62 million t, was to Italy. Considering container transport in 2021, Italy is the fourth country in container handling with 630,837 TEU after Greece, Israel and Egypt with more than 1 million TEU (General Directorate of Maritime Affairs of Turkey).

The current traffic level between the European and Asian sides causes enormous delays and difficulties for Istanbul. Unfortunately, there is an irregular port development in Turkey caused by the private port investments, considered a rapid response to container demand to compensate for the inadequate capacities of public ports. As a result, it can become detrimental to the efficient distribution of resources by causing overlapping investments [10].

Fig. 4 shows the amount of goods, which the Marmara, Aegean and Mediterranean Sea attracted between 1989 and 2016. There is a notable increase in the Marmara Sea comparing Aegean and Mediterranean ports. Turkey's busiest ports are concentrated around Istanbul and the Marmara Sea. Marmara is the region with the highest amount of cargo handled with 40% in 2020, followed by the Mediterranean region with 35% and the Aegean region with a

16% share. Therefore, it is vital to shift a part of this traffic to other ports of the country to support regional development [10].

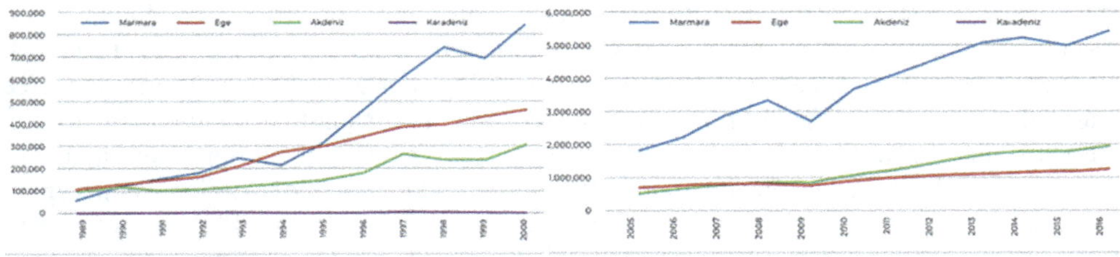

Figure 4: Container transport in Marmara (blue), Aegean (red) and Mediterranean (green) Sea in 1989–2000 and 2005–2016 (TEU) [10].

Italy is one of the first trade partners of Turkish ports and Turkey is the same for Italian ports, especially for port of Trieste. In this context, shifting some Turkish freight from Trieste to İzmir port in TEUs and using railways already connected with the hinterland serves this purpose. While freight transport by road over the years cannot shift to railway, it continues to take the least share, 4% in total and 1% international in 2020. International rail links by land are Bulgaria, Greece, Georgia, Armenia, Iran and Syria. Links with Armenia and Syria do not provide service in the current situation. There are several difficulties concerning international rail freight transportation mainly related to modernization works/constructions and security problems at the eastern border.

3 THE PORT OF TRIESTE

Trieste is a port city in north-eastern Italy and towards the end of a narrow strip of Italian territory lying between the Adriatic Sea, Slovenia and Croatia [11].

The port area is about 2.3 million m^2 (about 1.8 million m^2 of free zones). The length of the docks is 12 km with 58 berths. The maximum depth of water is 18 m. The total length of internal rail track is 70 km connected to the national and international network, aiming to serve shunting and/or assembling freight trains directly in the terminals. The port is in operation 24 hours/day, 365 days/year. The port operator is Adriafer Srl, company setup by the Trieste Port Authority. It is the only operator allowed at moving wagons and trains and organize the rail operations within the port. The Alpe Adria Spa is a multimodal transport operator that coordinates road, rail and sea carriers to organize and handle intermodal and combined transportation. The company runs a wide network of links between north-east Italy and central/eastern Europe. It has strongly developed in intermodal container transport and combined road–rail transport (Ro-La) [12], [13].

The port of Trieste is having the advantages of the special regime of free zones and a natural water depth up to 18 meters and ease of access for vessels. Intense road and rail links, regular ocean transportation services to/from China, India and the Far East takes competitive advantage in serving the markets of central and eastern Europe by saving more than 2,200 miles (more than four days sailing at 20 knots): Suez–Trieste distance is 1,300 miles, while Suez–north European ports distance is about 3,500 miles [13].

In 1719, the Austrian Emperor Carlo VI established The Free Port of Trieste. The 1947 Paris Peace Treaty and the Memorandum of London in 1954 gave an extra-territorial status and created the Free Port of Trieste's regime. Since then, customers can benefit from special conditions for import, export and transit operations, fiscal regime and customs procedures.

An EU Free Zone is not comparable to Trieste's Free Port because its regime is born from an international Treaty signed before the EU's existence. Thus, many of Trieste's Free Port benefits are not available in the EU Free zones [14]. Examples are free access granted to ships, trains and trucks of all flags, preferential maritime duties not chargable by Italian taxes, no time limit for freight storage, no customs formalities and duties, low harbour duties, simplified transit and custom systems in transit to outside of Italy; permitted free manipulation of goods (packaging, repackaging, labelling, sampling, eliminating brands, etc.) [15]. Foreign rail carriers can operate in the port of Trieste without limitations and non-EU trucks can be driven inside the Free Port area freely. If directed to a foreign country, they can reach the border (e.g. Austria or Slovenia) with a temporary permit. For instance, due to the limited number of annual transit permits provided to Turkey by the Italian government, the Trieste free status is a key attraction for Turkish operators as trucks are not subject to national quota restrictions [16].

Main terminals are Pier VII (TMT) which is a container terminal with the deepest natural draft of the Mediterranean Sea (18 m at the berth). Most of the traffic is currently concentrated at this terminal. On the five rail tracks of 600 m each, operate three rail-mounted stacking cranes able to work on up to five trains simultaneously. Pier VI (EMT) with drafts 9–10 m. The terminal has four rail tracks. The primary Ro-Ro connections leave from Pier VI of Trieste heading to Pendik and Ambarlı operated by EKOL. Pier V (SAMER/DFDS) is the main Ro-Ro terminal that can host three vessels simultaneously with a 12 m natural draft. It includes a renovated section devoted to rail operations featuring a rail-mounted stacking crane able to operate on four 310 m long tracks. This is the terminal with most of the Turkish freight traffic. Among other operators, SAMER is a specialist in ISU transport [11], [14].

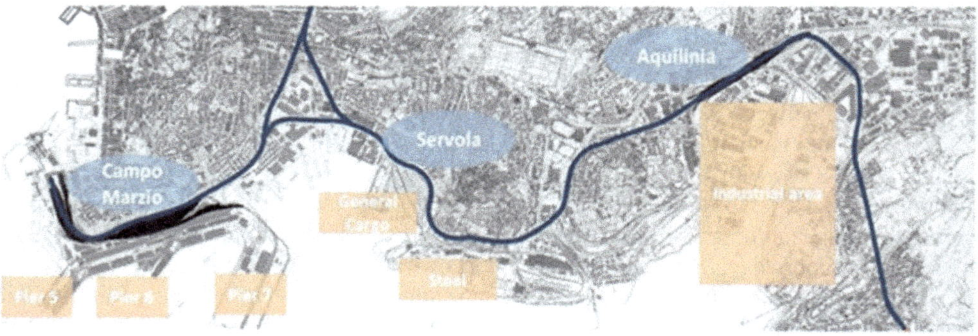

Figure 5: Structure of the port of Trieste [11].

The port of Trieste is in connection with to the national railway line, thus to the TEN-T networks, through the following nodes (Fig. 6): Campo Marzio, Servola and Aquilinia [11]. Concerning shunting operation in the terminals, trains have normally two sections because of the tracks' limited length (550 m). These two parts join again using the shunting locos on one of the 750 m long six tracks after waiting on dedicated sidings. Currently, two locomotives and two Zephyr vehicles are available for this shunting operation [12].

Connected to the Brennero railway line through Tarvisio and located close to rail lines directed to Ljubljana and Zagreb, Trieste stands out 52% of the full containers passing by train in 2017, while it was 30% in 2016 for 8,911 trains in 2019, according to AdriaFer data.

Figure 6: Railway infrastructure at port of Trieste: shunting area (green) and Campo Marzio station (RFI) (blue). *(Source: AdriaFer.)*

In the case of Ro-Ro traffic, the Turkey strategic relationship with Trieste created in the last decades represent the Mediterranean's busiest Motorway of the Sea. The mean value of sea distance is around 50% of land distance [17]. In 2020, approximately 60% of the Ro-Ro traffic of Trieste is linking Pendik, Yalova, Ambarlı, Mersin and Çeşme.

According to Torbianelli [18], during the early 1990s, when there was the Yugoslavian crisis in Balkan countries, Trieste–Istanbul relations became critical to maintaining the traffic. Considering the difficulties of motorway transportation passing through Serbia and Croatia, the maritime route got primary importance to prevent war [18].

In 2021, the port of Trieste ranked first among Italian ports for total throughput with about 55.4 million t. By looking at Fig. 7, the growing trend of port of Trieste is evident, in both t and TEUs. This trend requires upgrades, especially in the fields of capacity. Network's ability to withstand traffic increases of 15–20% each year is a challenging issue, makes mandatory planning infrastructural upgrades and uses the potential of new technologies and communication systems to optimise the use of existing infrastructures [19].

When growing intermodal demand concerned, the port faces several challenges related to efficiency and connections. Namely, upgrade of the railway infrastructures, optimizing existing infrastructures through ICT tools, and working with the other ports [5]. The railway traffic of the port has grown significantly and the traffic demand will keep growing in the following years according to the trend. Thus, the principal investments in infrastructures aim at increasing the total capacity for handling goods, with particular focus on railway capacity and intermodality [19].

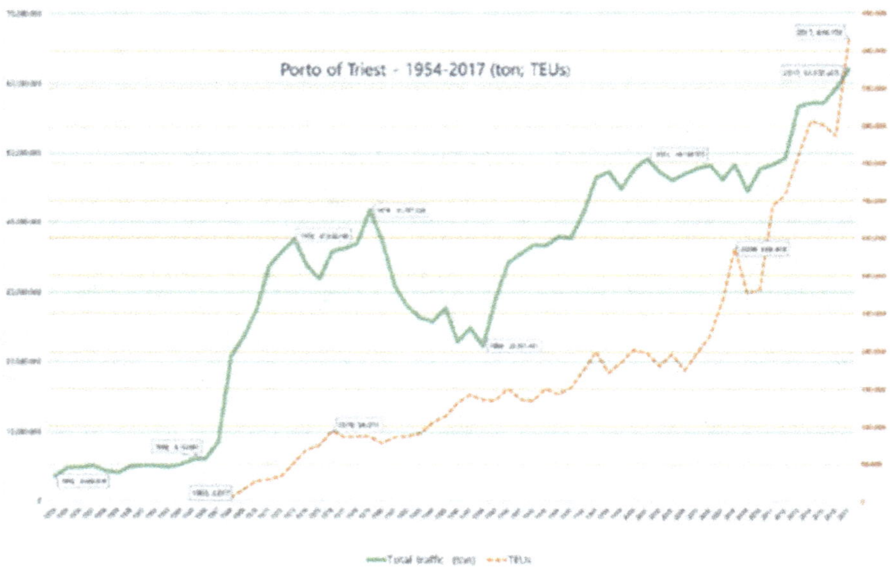

Figure 7: Traffic at port of Trieste 1954–2017 [20].

There are also challenges related to shunting operation, such as the not automated marshalling yard causing delays and posing higher risks to safety due to potential human errors, require renovation and optimisation [12], [19]. However, upgrading infrastructure is a long-term goal that expected to be completed within 5–10 years. Meanwhile, the port of Trieste aims to use IT solutions to increase the railway capacity by making administrative operations faster and automated [19]. In this context, the Global Project (2018–2025) Trieste RailPort, co-funded by the CEF Programme, consists of strengthening and adapting to significantly increasing trend the traffic at the port, which aims to further growth from about 10,000 up to 20,000 trains/year [20].

4 THE PORT OF İZMIR

İzmir is an old city at the west shore of Anatolia Peninsula and it was always an important port city (Fig. 8). Recently, it has been an important container port in Turkey for many years, owned and operated by the railway infrastructure manager TCDD. The capacity is 1,025,000 TEU, 4.6 million t of general cargo and dry bulk and 372,000 Ro-Ro vehicles per year. The depth of the water is 10–12 m at the approach channel and 6–11.5 m at the berths. Moreover, there is a difference of 2 cm between the designed depth and the existing depth due to shallowness. [2], [21], [22]. The distances by rail from İzmir to its hinterland are: 884 km to Istanbul, 824 km to Ankara, 1,167 km to Edirne, 133 km to Aydın, 67 km to Manisa and 264 km to Denizli.

In recent years, its total throughput has decreased dramatically after 2008 financial crises while its competitors, both in the Aegean and Marmara Region, have been developing their business rapidly. It could not recover its business after 2008 because of several factors mainly related to congestion, growing vessels sizes and new competitors [2], [21], [22]. In 2008, the port of İzmir reached the 891,701 TEU container throughput and it was the highest container handling port in Turkey. This number decreased over the years and had dropped to 517,376 TEU in 2021.

(a) (b)

Figure 8: (a) Rail network in the West of Turkey; and (b) Satellite image and existing rail infrastructure of port of İzmir. *(Source: TCDD.)*

There are several reasons for this loss but mainly related to the draft problem, its position in the city and connectivity of the port. The approach channel and berth depths are not adequate for the entry of big vessels. Moreover, the port located in the city centre that affects its connections quite negatively since the primary mode of transport is the road. According to 2021 data, port of İzmir had a 9.29% rail share (12.77% unloading 5.87% loading) in total 9,215,988 t and 53,549 wagons. The rail share was 18.74% in 2016 and this decrease is due to infrastructure and poor coordination of the railway network. The most significant impact is that the processing and delivery of cargo through the railway is highly inefficient due to delays, which affects the punctuality and the cost. Also, most of the port equipment has completed their economic life. The automation system, modernized in 2013, is a quite old equipment, incompatible with more modern automation systems [23]. Another reason is the privatization process caused, which brings the port of İzmir to face a severe cargo loss due to the inability to make the needed investments. This process took ten years until removed from privatization program but caused delays in dredging works as well as the ageing superstructure elements, then shifting the new investments to Marmara Sea [2].

5 SEA–RAIL TERMINAL DESIGN FOR PORT OF İZMIR

The rail terminal's general structure has transhipment tracks under a rail crane. The transhipment track is a rail track served by handling equipment. There are also waiting and driving tracks. If there is no space in transhipment, tracks or trains need to wait for departure to the existing intermodal network (Fig. 9). The driving track connect them to the network and the transfer points for yard trucks are at the side of transhipment tracks (Fig. 10) [24].

When a container ship calls to port and moors at the quay, containers onboard are subject to unloading and storage at the port until their forward by rail or road. The containers are unloaded into the marshalling area and wait to move to the storage areas or trucks for the transfer to the multimodal terminal by trucks to the intermodal terminal or directly to the hinterland [24], [25]. The number of transhipment tracks determines the maximum number of trains that can stay in the handling yard simultaneously. The operational capacity of shunting locomotives determines the number of operated trains. The connected railway marshalling station determines the shunting capacity [26]. The preferred access to terminal is organised from both ends; however, many terminals have dead-end tracks, such as Trieste and İzmir [27]. The capacity of a station defines as the maximum number of trains the station can process and perform, such as stops, overtaking manoeuvres and exit the station [26].

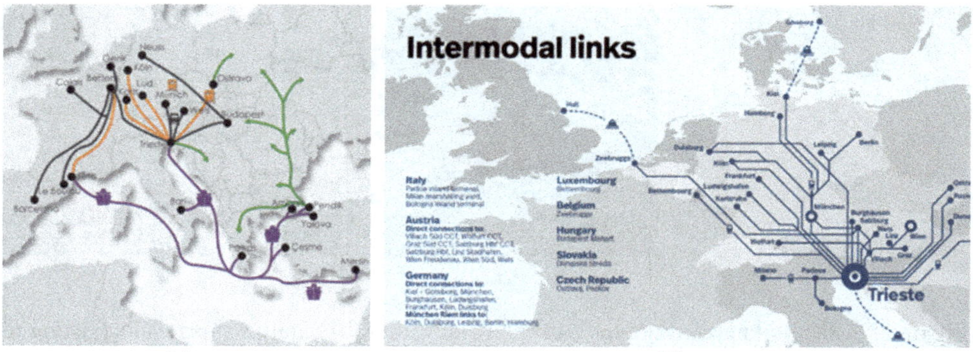

(a) (b)

Figure 9: (a) Intermodal connections of the port of Trieste [11]; and (b) Intermodal links of a major shipping company working between Turkey and Trieste. *(Source: EKOL.)*

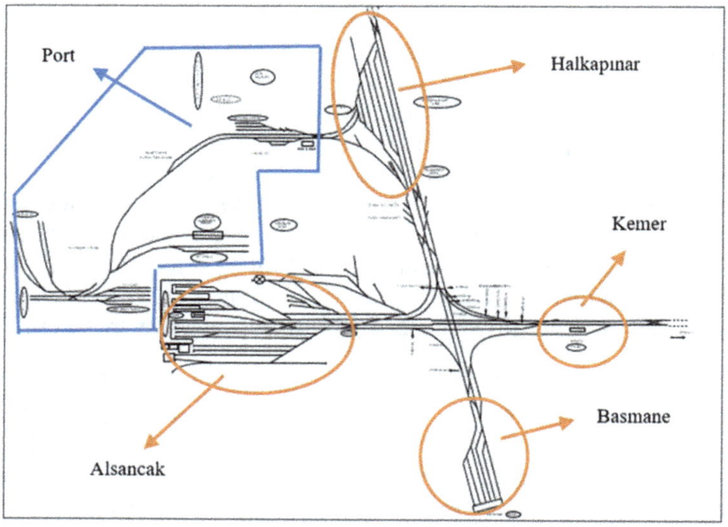

Figure 10: The existing rail infrastructure of port of İzmir and closest stations (Alsancak, Basmane, Halkapınar and Kemer). *(Source: TCDD.)*

The development of container sea–rail intermodal transport of port of İzmir is relatively backward comparing to its competitors (Fig. 11). This part of the paper will address a case study for designing a new rail terminal to increase the port's rail share, thus improving sea–rail intermodality. The design proposal includes analytical calculation, according to the transport requirements of the port, to estimate the design parameters: number of transhipment tracks and cranes and the traffic flow inside the rail terminal [24], [28].

The rail terminal capacity is 1,025,000 TEU/year. Aiming at 50% rail share, the total throughput requirement of the rail terminal is 512,500 TEU/year. Table 1 indicates the

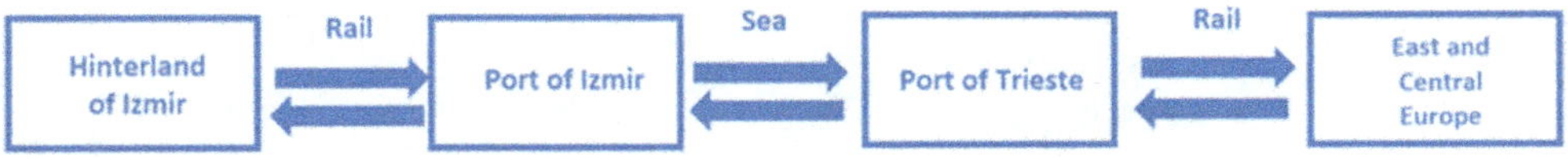

Figure 11: The suggested transportation model.

Table 1: Formulas and constant parameters for determining the values of design parameters.

Notations	Description
$Avan$	Effective handling capacity of a crane per hour $Avan = Tvan \times \eta T$
CV	Transport requirement in the number of containers (van) $CV = \frac{CTEU}{TEUf}$
Cea	The number of cranes required (design parameter) $Cea = \frac{CV \times Fp}{Avan \times Wd \times Wy}$
YHC	Handling capacity of cranes per year $YHC = Avan \times Wd \times Wy \times Cea$
DHC	The number of loading and unloading containers per day $DHC = \frac{CV \times Fp}{Wy}$
THC	Handling capacity of a crane per day $THC = Avan \times Wd$
Chf	The average number of handlings per container $Chf = DL + \{(1 - DL) \times 2\}$
$Trteu$	The loading capacity per train (TEU) $Trteu = NR \times 2(TEU)$
$Trvan$	The loading capacity per train in the number containers $Travn = \frac{Trteu}{TEUf}$
$Trhc$	The handling capacity of a crane (in TEU) per hour $Trhc = TEUf \times Avan$
$Trht$	Loading and unloading time per train by a crane $Trht = Trteu / Trhc$
$Trtot$	Total service time per train by a crane $Trtot = Trht + Trot$
$Ddtn$	The number of trains per day that arrive at the rail terminal $Ddtn = \frac{DHC}{Trvan}$
Dtn	The number of assigned trains per day per track $Dtn = \frac{Wd}{Trtot}$
$Dntn$	The required number of transshipment tracks (design parameter) $Dntn = \frac{Ddtn}{Dtn}$
Ct	Theoretical cycle time of the crane
$Tvan$	The handling capacity of the crane per hour $(1/C_t)$
ηT	The utilization of the crane
Wy	The number of workdays per year
Wd	Working hours per day
$TEUf$	TEU factor to convert the number of containers to TEU
$CTEU$	Transport requirement per year (TEU)
Fp	Peak factor
DL	Ratio of the direct transportation
NR	The average number of wagons per train
$Trot$	Time for the locomotive operations

Table 2: Input values of constant parameters [25].

Notations	Input values
Ct	82.3 s
$Tvan$	43.5 TEU/h
ηT	0.45
Wy	365 day/year
Wd	24 h/day
$TEUf$	1.48
$CTEU$	512,500 TEU/year
Fp	1.25
DL	0.6
NR	20

Table 3: Design parameters of the rail terminal.

Notations	Output values
Cea	3 cranes
YHC	517,716 TEU/year
DHC	1,186 TEU/day
$Ddtn$	44 trains/day
Dtn	10 trains/day/track
$Dntn$	5 tracks

numerical formulas for estimating the values of design parameters. Table 2 includes the input values of constant parameters. Table 3 summarise the numerical results and Fig. 12 the layout of the transhipment terminal. According to this analytical approach, the rail terminal needs a

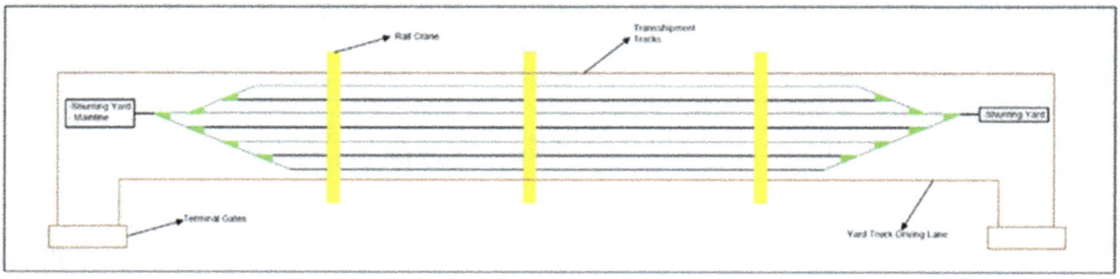

Figure 12: The layout of proposed sea–rail intermodal terminal for port of İzmir.

minimum of three rail cranes and five transhipment tracks to meet the planned capacity and performance. Two more transhipment tracks would allow at possible future growth of the handling requirement and holding. There are two gates for yard trucks to enter and, because of insufficient areas for standard 750 m tracks, the designed layout composed of 550 m tracks.

According to the TCDD network data, the capacity of Halkapınar-İzmir port is 40 double trains with double track, signalized at 3 km distance, the capacity of Basmane-Kemer stations is 40 double trains with single track, signalized at 1 km distance.

6 CONCLUSIONS

This paper underlined how important sea–rail intermodality is for countries with peninsular morphology, like Turkey and Italy. Nevertheless, looking at the rail modal shares in sea–rail intermodality, it is self-evident that Turkey needs to invest in the rail connections of the ports. Analysing port of Trieste was also crucial for several reasons. First, it is the port with the highest rail share in Italy and has been highly invested for this purpose. Secondly, it receives the highest amount of Turkish maritime freight, mainly directed to Europe. Almost all of them is coming from the Marmara. Moreover, almost all the freight is arriving at the Marmara ports by road. The suggested solution in this study is receiving containers from port of Trieste and then directed to the hinterland by rail from the rail terminal of port of İzmir and vice versa. With realizing this target, trade between ports of İzmir and Trieste is affordable with 50% by rail from both sides. Because a rail modal share under 10% is not acceptable for TCDD.

To increase the rail modal share of port of İzmir, the aim was to achieve the rail terminal dimensioning by an analytical approach. Consequently, constructing a rail terminal with three rail gantry cranes and seven transhipment tracks will make it possible to have a 50% rail share. For further research, the capacity of storage and parking areas, shunting and station capacity of closest stations to reach hinterland (e.g., Halkapınar and Basmane) as well as the technology of gantry cranes need to be determined because these factors also influence rail performance and are difficult to reverse.

Finally, the reason of planned investments for port of Trieste was its rapidly increasing demand. In contrast, the necessary investments in port of İzmir are due to its dramatic cargo loss.

REFERENCES

[1] UTİKAD, Lojistik Sektörü Raporu, 2020.
[2] İzmir Kalkınma Ajansı, TCDD İzmir Alsancak Limanı'nın Geçmişten günümüze bölge ekonomisi açısından değerlendirilmesi, 2019.

[3] Pastori, E., Modal share of freight transport to and from EU ports. European Parliament, Directorate-General for Internal Policies, 2015.

[4] Lupi, M., Pratelli, A., Licandro, C. & Farina, A., The evolution of deep-sea container routes: The Italian case. *Transport Problems*, **14**, pp. 69–80, 2019. DOI: 10.21307/tp.2019.14.1.7.

[5] SRM, Italian maritime economy: New scenarios in the Mediterranean: Suez and China, the strategies of big carriers, new technologies and energy routes, pp. 1–10, 2019.

[6] Invest in Italy, Italian ports and freight villages: Investment opportunities, Silk Road Exhibition, 2019.

[7] İzmir Ticaret Odası, İtalya Cumhuriyeti Ülke Raporu, 2021.

[8] ICID, Irrigation and drainage in the World: A global review, Italy, pp. 1–19.

[9] Lupi, M., Pratelli, A., Giachetti, A. & Farina, A., Rail freight transport in Italy: An analysis of combined transport connections. *IF Ingegneria Ferroviaria*, **3**, pp. 209–245, 2018.

[10] Liman Gelişim Stratejisi, 2020.

[11] Halatsis, A., Integrated and Sustainable Transport in Efficient Network (ISTEN), pp. 1–28, 2018.

[12] Licciardello, R., Adamko, N., Deleplanque, S., Hosteins, P., Liu, R., Pellegrini, P., Peterson, A., Wahlborg, M. & Zatko, M., Integrating yards, network, and optimisation models towards real-time rail freight yard operations. *IF Ingegneria Ferroviaria*, **6**, pp. 417–440, 2020.

[13] Trieste Port Authority, The Port of Trieste, p. 66, 2012.

[14] ASPT, The Free Port of Trieste, infrastructure and logistic network, 2019.

[15] https://www.porto.trieste.it/.

[16] Gregori, M., The Free Port of Trieste: An analysis of the current legal framework, maritime, port and transport law between legacies of the past and modernization, 2018.

[17] Malavasi, G. & Ricci, S., The sea-side port capacity: A synthetic evaluation model. *WIT Transactions on the Built Environment*, WIT Press: Southampton and Boston, pp. 471–480, 2005.

[18] Torbianelli, V.A., Turkish ro-ro traffic in the port of Trieste. *Integrated Transport Review*, **6**(11), pp. 317–321, 1999.

[19] Interreg SMARTLOGI, Studi sui bisogni territoriali per il Porto di Trieste e il FVG, 2018.

[20] Interreg TALKNET, Action plan to improve multimodal nodes efficiency and connections, Trieste (NAPA), p. 1, 2020.

[21] https://www.tcdd.gov.tr/kurumsal/izmir-limani.

[22] Alpcan, O., Efficiency and competitiveness analysis of port of İzmir. World Maritime University, Master of Science in Maritime Affairs, 2019.

[23] Yormaz, D. & Oral, E.Z., TCDD İzmir Limanı Rıhtım Ötelemeleri, 2014.

[24] Lee, B.K., Jung, B.J., Kim, K.H., Park, S.O. & Seo, J.H., A simulation study for designing a rail terminal in a container port. *Proceedings of the 2006 Winter Simulation Conference*, pp. 1388–1397, 2006. DOI: 10.1109/WSC.2006.323239.

[25] Vis, I.F.A. & De Koster, R., Transhipment of containers at a container terminal: An overview. *European Journal of Operational Research*, **147**, pp. 1–16, 2003.

[26] Marinov, M., Şahin, I., Ricci, S. & Vasic Franklin, G., Railway operations, timetabling and control. *Research in Transportation Economics*, **41**(1), pp. 59–75, 2013. DOI: 10.1016/j.retrec.2012.10.003.

[27] Ballis, A. & Golias, J., Comparative evaluation of existing and innovative rail–road freight transport terminals. *Transportation Research Part A: Policy and Practice*, **36**, pp. 593–611, 2002.
[28] Sun, X. & Luo, B., A study on the development of container sea–rail intermodal transport of Xiamen port in the context of national logistics hub. *Conference Series: Earth and Environmental Science*, Institute of Physics Publishing, 2020.

METHODOLOGY FRAMEWORK FOR PRIORITISATION OF RENEWABLE ENERGY SOURCES IN PORT AREAS

SOFIA AGOSTINELLI[1], MEHDI NESHAT[2], MEYSAM MAJIDI NEZHAD[1], GIUSEPPE PIRAS[1] & DAVIDE ASTIASO GARCIA[3]
[1]Department of Astronautics, Electrical and Energy Engineering, Sapienza University of Rome, Italy
[2]Center for Artificial Intelligence Research and Optimization, Torrens University Australia, Australia
[3]Department of Planning, Design, and Technology of Architecture, Sapienza University of Rome, Italy

ABSTRACT

Ports play a crucial role in increasing the decarbonisation of urban environments to mitigate the environmental impacts of maritime transport and promote sustainable intermodal mobility. Various efforts have been made to increase energy self-sufficiency using renewable energy sources (RESs) in different ports worldwide. However, the ports played an essential role in the pollution process of the nearest cities due to the short distance and merging with urban areas. In this case, solar and wind were measured using the Modern-Era Retrospective analysis for Research and Applications, Version 2 (MERRA-2) data of four Lazio province ports. Each RES was evaluated using 10 years of monthly data for mapping and 1 year of hourly data for potential assessment and energy converters installation. Furthermore, the time series method has been considered to design and develop better management of RESs for decision making monitoring the energy needs of ports. This time series method has been applied to the generated energy source based on various parameters of the RESs used in port.
Keywords: renewable energy sources, decision making, geographic information system, MERRA-2, time series analysis.

1 INTRODUCTION

The Paris Agreement on climate change clarifies that significant step should be taken towards a 100% decarbonisation by 2050 [1]. The total electricity generating capacity for industrial ports could increase more than tenfold thanks to renewable energy sources (RESs) by 2050. However, current activities in the port are significant contributors to CO_2 emissions and require drastic rethinking. Ports present a prominent energy source as crucial locations for energy production from RESs: an awareness that inspired the recent European Green Deal (EGD), in which ports are conceived as possible energy poles and engines of the blue economy next generation EU Program. These considerations are the main impetus for this kind of study, aiming to present efficient methods and models in European ports of technology to improve the energy efficiency of port areas. A port can provide RESs and support vessels with clean electricity. Different approaches of alternative energy for ports were proposed, such as Antwerp, Genoa, Venice, San Diego [1].

Some studies were carried out to evaluate the applying of RESs for the decarbonisation of the Italian ports, for example applying wind and solar energy [2], [3] in the European ports [4]–[6]. Of course, reducing greenhouse gas (GHG) and pollutant emissions are directly linked with energy efficiency interventions, electrification of equipment, and the use of RES or alternative fuels. These aspects, along with increasing operational efficiency, are crucial for increasing ports' energy and environmental sustainability in the next generation. Ports can be classified into six types by considering various factors such as location, water depth, purpose and size of ships [3]: (a) domestic ports, (b) fishing ports, (c) hot water ports, (d) dry ports, (e) seaports, and (f) cargo ports. Each cargo port has three functional areas, namely quayside, yard side and landside. The study is aimed to design, develop, test and validate a replicable multidisciplinary approach that includes method for identifying and quantifying

WIT Transactions on The Built Environment, Vol 212, © 2022 WIT Press
www.witpress.com, ISSN 1743-3509 (on-line)
doi:10.2495/UMT220101

the potential of RESs interventions in ports. Finally, a method has been designed and developed to enhance network intelligence to better manage RESs for decision making monitoring the energy needs of ports. This study uses reanalysis data and dataset has been designed, developed, and launched by the NASA, which has high datasets. These reanalysis dataset source has been used for discovering, extracting, processing and mapping of RESs potential analysis to better understand ports as a preferred area with very high time resolution.

2 CASE STUDY

In this article we have studied the four ports of Lazio: Anzio (41°26′52.61″N 12°37′44.59″E) is a city and commune on the coast of the Lazio region about 51 km south of Rome. Formia (41°16′N 13°37′E) is situated on the Gulf of Gaeta on the west coast of Lazio. Formia port is a general cargo port formed by a breakwater and a jetty, accommodating fishing vessels only. Terracina (41°17′N 13°15′E) is an Italian city and located on the coast 56 km southeast of Rome. Ventotene (40°47′N 13°25′E) is one of the Pontine Islands in the Tyrrhenian Sea.

3 MATERIAL AND METHODS

In this paper, wind speed and solar radiations potential of the Lazio province ports have been analyzed using more than ten years of monthly data for mapping and one year of hourly data for time series analysis from the MERRA-2 reanalysis dataset. This long-term analysis has been done to evaluate and identify different ports with the exciting potential of Lazio ports' wind speed and solar radiations for micro wind turbines and PV installations. Furthermore, the priority decision-making of suitable potential ports is examined based on the following parameters.

3.1 Long term wind speed and solar irradiation mapping

Firstly, more than 10 years (2010–2021) of monthly data from the MERRA-2 reanalysis dataset have been used to evaluate and identify wind speed and solar irradiation potential classification ports. This data has been used to better understand the wind speed and solar radiation of Lazio ports to install micro wind turbines and PV facilities in one interval.

3.2 Port's analysis using hourly wind speed and solar irradiation data

Secondly, 1 year of hourly data from the MERRA-2 reanalysis dataset has been used to monitor and assess wind speed and solar potential time series analysis of ports areas with exciting potential to better understand the wind speed and solar irradiation assessment wind turbines and PV installations and make better decision-making. Furthermore, this year (2020–2021) hourly data from the MERRA-2 reanalysis dataset relative with ports areas outputs has been used as input of the next step, which means 3.3.

3.3 Statistical analysis of wind speed

In order to design and evaluate wind energy systems, wind speed parameters play a significant role. The interval of the recorded wind speed time series is hourly were acquired from the MERRA-2 in Lazio port across 12 months period from 2020 to 2021. In this study, we apply a Pearson's linear correlation coefficient test in order to analyse the level of correlation between the wind speed data of four ports. The correlation coefficient of the population between two sets generated randomly; X and Y is defined as follows:

$$px,y = \text{corr (x,y)} = cov(x,y)/\sigma_x\sigma_y \ = E[(X-\mu_X)(Y-\ \mu_Y)]/\sigma_X\,\sigma_Y, \tag{1}$$

where μ_X and μ_Y are the expected values of X and Y and both σ_X and σ_Y are the standard deviations. First and foremost, observation is a positive association among all ports in terms of wind speed data. Secondly, the highest positive correlation can be noticed between Ventotene and Formia port. In the third step of the wind speed data statistical analysis, an autocorrelation test is used. In order to define an autocorrelation coefficient lag k (r_k) for a sequential time series $s_1, s_{2,...,} s_k$ as follows:

$$r_k = \frac{1}{T}\sum_{t=1}^{T-k}(S_t - \bar{S})(S_{t+k} - \bar{S})/c_0, \quad \text{k} = 0,1, \ldots, \text{L}, \tag{2}$$

where c_0 denotes the sample variance of the sequential time series data. T is the effective sample size of S_t, and also $\bar{S}$ is the average of time series S_t. Generally, $L = \frac{T}{4}$ and depends on the time series data length.

Indeed, autocorrelation regards the correlation strength of the corresponding variables between two consecutive time intervals. This analysis is generally applied to indicate that how the lagged version of the variable values is described as the primary variant of it in sequential data. The standard range of autocorrelation is between –1 and 1. A negative autocorrelation is imposed where the value is from –1 to 0. Inversely, we can see a positive autocorrelation with a value between 0 and 1 [7].

4 RESULTS

4.1 Long term wind speed assessment and mapping

In the Lazio ports, the highest wind speed values can observe in the Ventotene port near 5.5 m/s. The Anzio and the Terracina port with wind speeds of more than 5 and 4.5 m/s, also Formia port areas have wind speeds of about 4 m/s in the same period (Fig. 1). Fig. 1 shows wind speed analyzed using the MERRA-2 reanalysis dataset for the four case studies of Lazio and Fig. 2(a) and (b) shows a monthly and seasonality wind speed using hourly time series from 2020 to 2021 for the four Lazio ports.

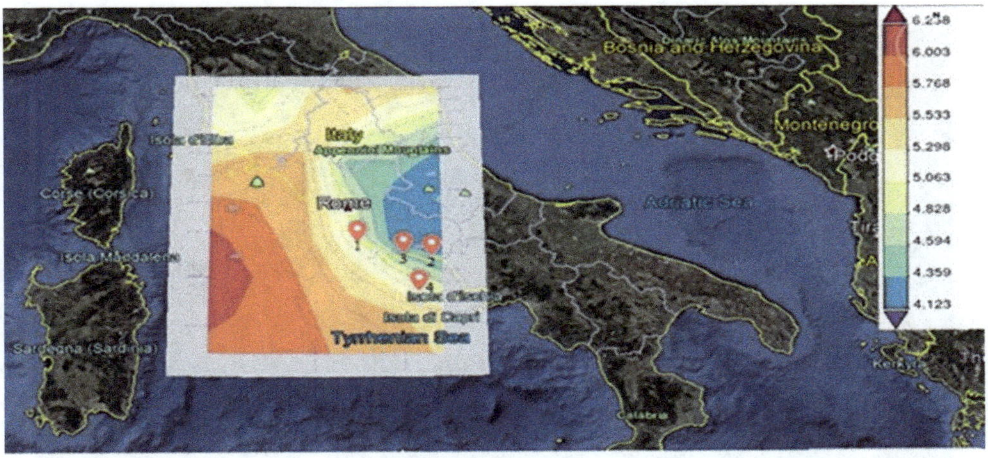

Figure 1: Surface wind speed monthly 0.5 × 0.625 deg. m s^{-1} from May 2011 to March 2021. Region 10.0707E, 39.6929N, 14.3334E, 43.648N.

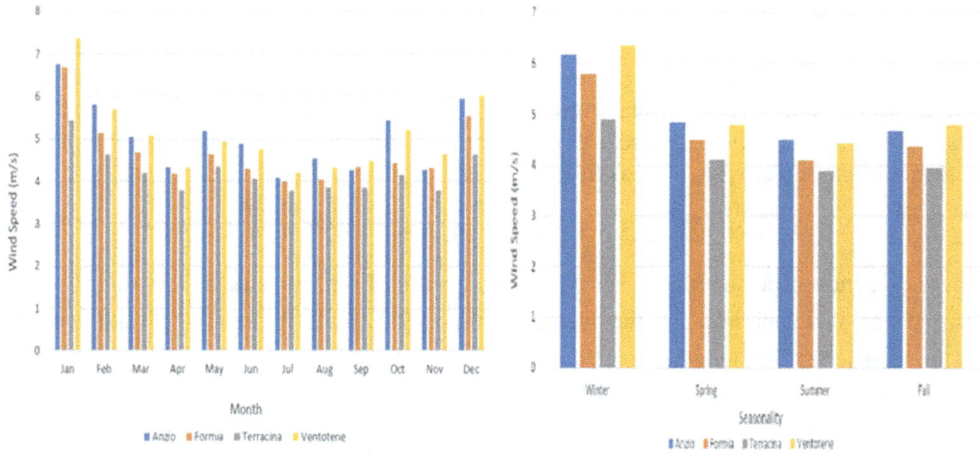

(a) (b)

Figure 2: (a) Monthly; and (b) Seasonality wind speed using hourly time series from 2020 to 2021.

4.2 Long term solar radiation assessment and mapping

In the Lazio ports, the highest solar radiation values can observe in the Ventotene port near 3.169 W m^{-2}. The Anzio and the Terracina port with solar radiation of more than 3.2 and 3 W m^{-2} (between 2011 to 2021), also several Formia port areas have solar radiation of about 2.5 W m^{-2} in the same period (Fig. 3). Fig. 3 shows solar radiation analyzed using the MERRA-2 reanalysis dataset for the four case studies of Lazio. Fig. 4(a) and (b) shows a monthly and seasonality solar radiation using hourly time series from 2020 to 2021 the four Lazio ports.

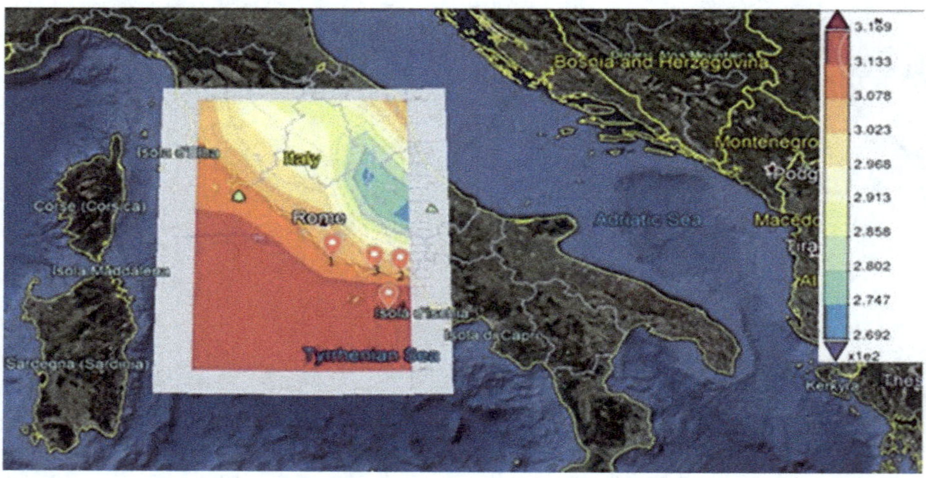

Figure 3: Surface absorbed longwave radiation assuming clear sky monthly 0.5 × 0.625 deg. W m^{-2} from May 2011 to March 2021. Region 10.0707E, 39.6929N, 14.3334E, 43.648N.

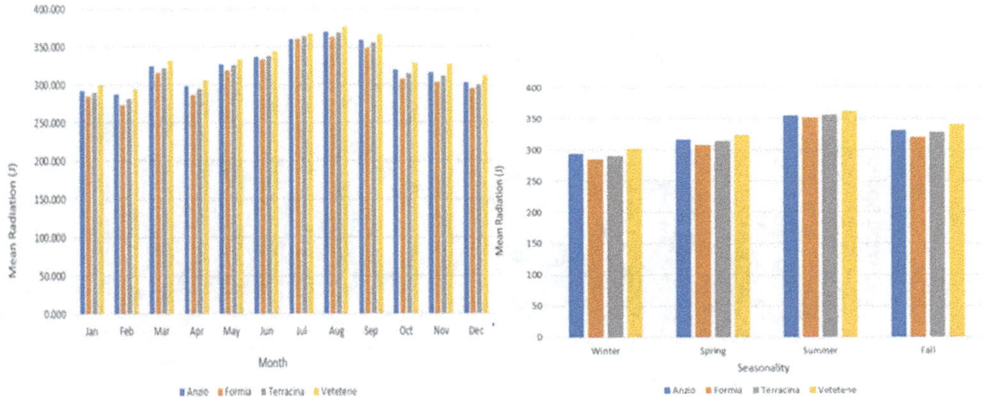

Figure 4: Monthly and seasonality solar radiation using hourly time series from 2020 to 2021.

4.3 Statistical analysis of wind speed

In the first step, to present and compare the wind characteristics at each particular ports, we present the stochastic behaviour of the recorded wind speed in Fig. 5.

Three different time resolutions (hourly, daily and monthly) are applied to compare the detailed relationship among the four ports and can be seen in Fig. 5(a)–(c), respectively. From Fig. 5(c), the principal observation is that the general models of wind speed in four ports are relatively comparable. However, as the details of wind direction data are not available and have a direct influence on the performance of the forecasting model, it is hard to compare the wind pattern in these locations. The frequency, mean, and standard deviation of 1-year wind speed values calculated for the available time-series data for four ports, and also are exhibited in Fig. 6(a)–(d). Furthermore, the probability distribution fitting is applied in order to fit a normal probability distribution to the wind speed time series regarding the repeated measurement of the variable wind data. The distribution fitting aims to predict the probability or the occurrence frequency of the wind speed magnitude in a certain interval [8]. It can be seen that the minimum and maximum average wind speed recorded is related to Terracina and Ventotene port at 4.175 and 5.065 (m/s), respectively.

In the following, we performed the correlation coefficient test that is an indicator of the linear relationship strength between both various variables. Where this coefficient is greater than zero designates a positive relationship between two applied time-series data. On the other hand, a rate that is less than zero implies an inverse or negative relationship. Conclusively, when the correlation coefficient is equal to zero, it shows no relationship between the two variables.

Fig. 7 shows the scatter, histogram and correlation coefficients of each pair location. The autocorrelation function of the measured wind data at four ports can be seen in Fig. 8. Interestingly, in all locations, there is the same pattern of the autocorrelation function. It shows that the wind speeds are strongly autocorrelated at the initial 15 lags. In the following, we can observe that less fully autocorrelated are at longer lags, and in most cases, the autocorrelation tends to the negative values. Another important point is that there is almost no daily pattern among the observed wind speed data of the ports.

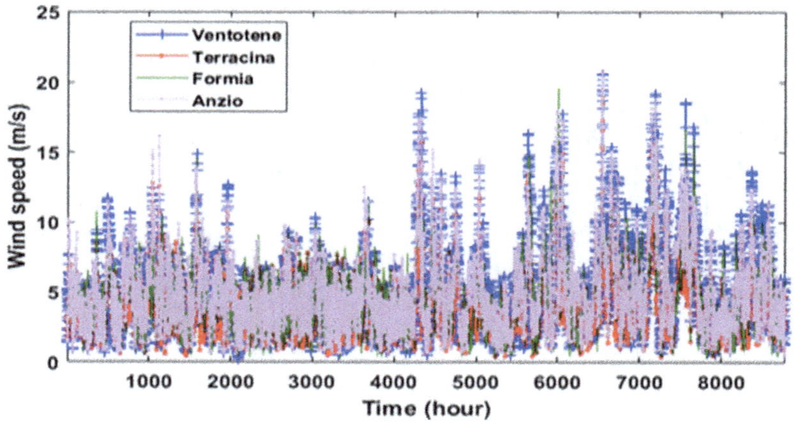

(a)

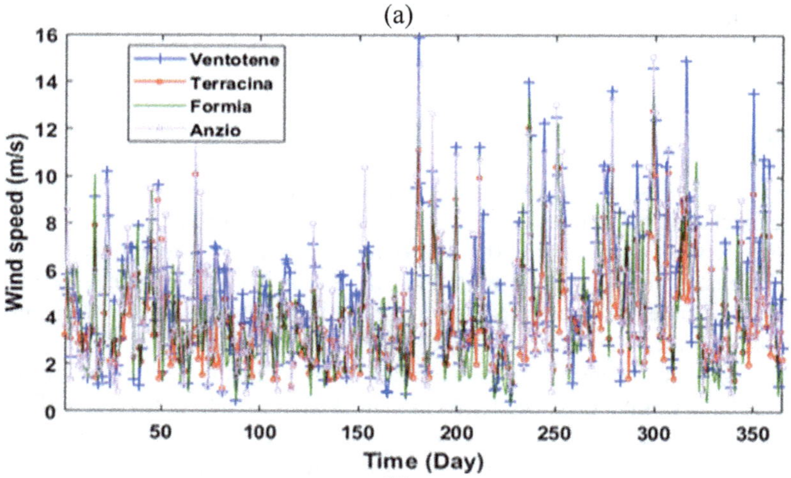

(b)

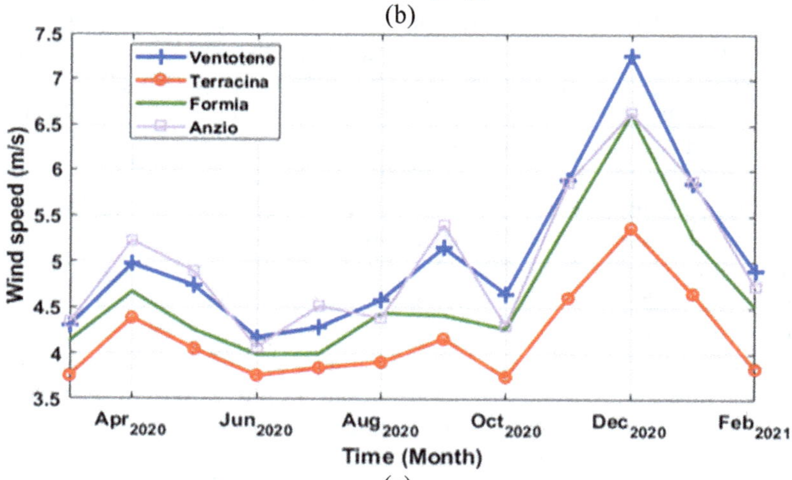

(c)

Figure 5: Sequential wind centre, Sapienza Lazio ports: Ventotene, Terracina, Formia and Anzio. (a) Recorded per hour; (b) Recorded per day; and (c) Recorded per month.

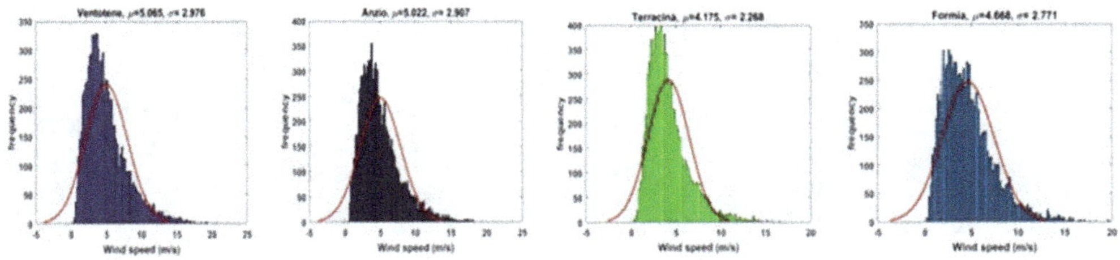

(a) (b) (c) (d)

Figure 6: The histogram with a normal distribution fit of the wind speed data for Lazio ports. (a) Ventotene; (b) Anzio; (c) Terracina; and (d) Formia. The estimated fittest normal distribution parameters: mean and variance are shown per each port.

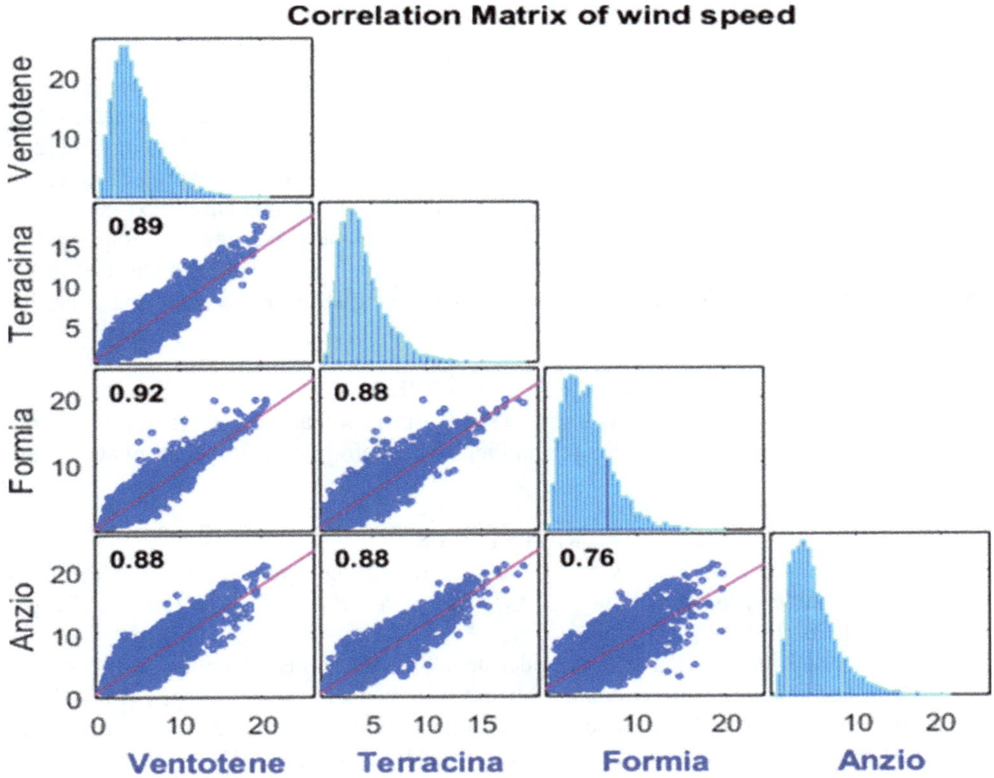

Figure 7: The Pearson's linear correlation coefficients of pairs ports wind speed data are depicted using scatter plots of variable pairs (Ventotene, Anzio, Terracina, and Formia) in the off diagonal. The slopes of the lines in the scatter plots notify the represented correlation coefficients

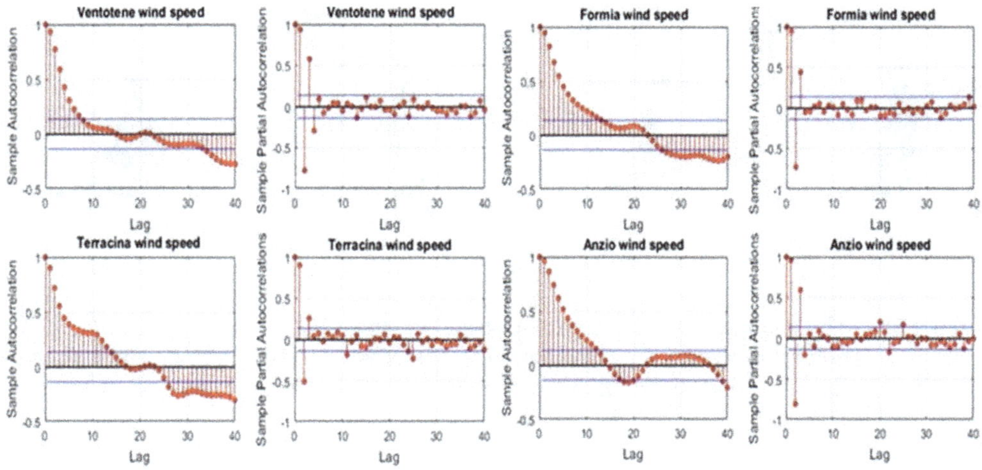

Figure 8: The sample autocorrelation function (ACF) and partial autocorrelation of the univariate, stochastic wind speed time series with confidence bounds for four ports (Ventotene, Anzio, Terracina, and Formia).

5 CONCLUSIONS

In recent years, due to the rapid growth of science and technology on RESs analysis in ports, there are still significant research gaps that need to be addressed in future research. RESs analysis is one of the hot topics of research that advances existing technology directly affecting future research prospects. Investing and developing RESs is a big project, and such investments require careful statistical and economic analysis. In this case, next-generation ports use automation, power and intelligent energy management systems. It is important to note that further research in the early stages must be of high data quality to succeed in the analysis. Managing sustainable energy development using RESs is an emerging issue for ports.

ACKNOWLEDGEMENT

This research has been carried out within ODYSSEA project that received funding from the European Union's Horizon 2020 research and innovation programme under grant agreement No 727277.

REFERENCES

[1] Bjerkan, K.Y. & Seter, H., Reviewing tools and technologies for sustainable ports: Does research enable decision making in ports? *Transp. Res. Part D Transp. Environ.*, **72**, pp. 243–260, 2019.

[2] Kang, D. & Kim, S., Conceptual model development of sustainability practices: The case of port operations for collaboration and governance. *Sustain.*, **9**, pp. 1–15, 2017.

[3] Badurina, P., Cukrov, M. & Dundović, Č., Contribution to the implementation of "Green Port" concept in Croatian seaports. *Pomorstvo*, **31**, pp. 10–17, 2017.

[4] Lam, J.S.L. & Notteboom, T., The greening of ports: A comparison of port management tools used by leading ports in Asia and Europe. *Transp. Rev.*, **34**, pp. 169–189, 2014.

[5] Maleki, A., Optimal operation of a grid-connected fuel cell based combined heat and power systems using particle swarm optimisation for residential sector. *Int. J. Ambient Energy*, **42**, pp. 550–557, 2021.
[6] Acciaro, M., Ghiara, H. & Cusano, M.I., Energy management in seaports: A new role for port authorities. *Energy Policy*, **71**, pp. 4–12, 2014. DOI: 10.1016/j.enpol.2014.04.013.
[7] Margrét Jónsdóttir, G. & Milano, F., Data-based continuous wind speed models with arbitrary probability distribution and autocorrelation. *Renewable Energy*, **143**, pp. 368–376, 2019.
[8] Zhang, S., Chen, Y., Xiao, J., Zhang, W. & Feng, R., Hybrid wind speed forecasting model based on multivariate data secondary decomposition approach and deep learning algorithm with attention mechanism. *Renewable Energy*, **174**, pp. 688–704, 2021.

EMERGING ICT AND PORT COMMUNITY SYSTEMS: A SURVEY OF SCIENTIFIC LITERATURE

GIUSEPPE MUSOLINO[1], GIROLAMO PEDA[2] & FRANCESCO RUSSO[1]
[1]DIIES – Dipartimento di Ingegneria dell'Informazione, delle Infrastrutture e dell'Energia Sostenibile, Università Mediterranea di Reggio Calabria, Italy
[2]Department of Engineering, Università Mediterranea di Reggio Calabria, Italy

ABSTRACT

Emerging information and communication technologies (ICTs) were identified as the main pillar of the fifth-generation ports. Ports' issues were solved with stand-alone ICT solutions adopted by each stakeholder. Today new benefits and challenges are connected with the introduction of shared emerging ICTs among port stakeholders, generating efficiencies in the solution of shared decisions. One of the tools that supports the interaction between port stakeholders is the Port Community System (PCS), which has been the object of numerous studies in recent years aimed at defining its benefits in terms of reducing the costs, or disutilities, of port operations. The paper presents a state of the art relating the Port Community Systems (PCS), in which a classification of scientific papers selected from the scientific databases present on the web is reported according to some classification criteria. The results of this analysis show that in the last decade the study and implementation, even experimental, of the PCSs inside ports has undergone a relevant acceleration. This is due to the fact that for port communities, being equipped with the PCS is now of strategic importance since its presence increases the competitiveness of the whole port. The work is useful both for port planners and for technicians of port manufacturing equipment companies, because it allows to identify the potential advantages obtained with the introduction of advanced evolutions of PCS.
Keywords: fifth-generation port, port community systems (PCSs), emerging ICT, IoT.

1 INTRODUCTION

Since ancient times ports are the gate for the exchange of freight and people. The evolution of ports may be described according to the concept of generation (see [1]–[3] and the references included). The first-generation of ports was built close to the cities, and it was the dominant model for centuries [4], [5]. The second-generation of ports was built close to industrial areas to support the supply of raw materials and final products of industrial plants in the second half of the 20th century [6], [7]. The third-generation of ports emerged after 1980s, principally due to worldwide diffusion of container and to the growing requirements of the international trade, becoming generators of added value. A set of publications presents the structural factors of a European container hub port in transition towards a third-generation. The limits and weaknesses of the port area and the general actions to reduce the port costs were identified in Russo and Chilà [8], [9]. The external structural factors of the port generated by the relationship with the hinterland and with research and development centers have been studied respectively in Russo et al. [10] and Russo and Rindone [11]. The internal structural factors of the port in relation to three industrial sectors (logistics, agri-food and mechanics) were analyzed in Musolino and Chilà [12], Musolino and Rosaria Trecozzi [13] and Musolino et al. [14]. The introduction of a SEZ in a third-generation port was described in the following publications. The publications by Russo et al. [15] and Musolino et al. [16] present respectively the modelling framework and the aggregate economic impact of a SEZ on the port and its hinterland. The territorial attractiveness and the urban planning connected with the activation of SEZ is presented in Musolino and Panuccio [17], while the impact of SEZ on the system of higher education and research is described in Rindone et al.

www.witpress.com, ISSN 1743-3509 (on-line)
doi:10.2495/UMT220111

[18]. The study by Pellicanò and Trecozzi [19] shows a comparison of times connected to the export/import of goods for European countries having similar technical-administrative structure. The fourth-generation of ports raised in the last two decades, when ports adopted a cooperative attitude with closer ports (see Russo and Musolino [20] and references included). Today fifth generation of ports is characterized by the use of emerging information and communication technologies (ICT) in the interaction between port stakeholders. The introduction of emerging ICT inside ports has shown that ICT is a further pillar of port competitiveness alongside the traditional ones as the port closeness to the cities or to the industrial plants, and the capacity to generate added value, the dotation of infrastructures and shared services (see Russo and Musolino [21], [22] and references included).

Financial and information transactions inside ports are supported by the adoption of stand-alone ICT systems. They increase the efficiency of the individual port operations associated the stakeholders, but they do not increase the efficiency of port operations that imply shared decisions between the stakeholders. Shared decisions are relevant in the three main flows present in the port: cargo, information and financial. In terms of cargo flow, cargo management involves the exchange of data connected to port operations (e.g. dock and yard operations). By considering the thousands of handled containers on ships, the complexity increases exponentially, and the traditional approach to container management generates a low degree of capacity utilization of ships and port infrastructures.

The dissemination of potentialities offered by emerging ICT was made possible by the capabilities of Port Community Systems (PCSs). The role of emerging ICT in the ports was presented in Russo and Musolino [21], in terms of assessment of stakeholders' utilities in sharing decisions about port operations. Some case studies of emerging ICT in ports operations are reported in Russo and Musolino [22], with a focus on internet-of-things. The impact of PCS characteristics on port performance is presented in Caldeirinha et al. [23], while a structured literature review on PCSs through the definition of the "wave" concept is reported in Moros-Daza et al. [24]. Finally, in Carlan et al. [25], it was developed a cost–benefit framework to assess the impacts of PCS on port competitiveness. PCSs aim to increase cooperation between port stakeholders, and consequently to increase their competitiveness compared to other ports. Cooperation is possible by integrating procedures, and by sharing of information and documents.

In the above context, the paper presents a survey regarding the scientific literature on PCSs and regarding the interactions that PCSs could have with emerging ICT. In particular, the interactions between PCSs and the Internet of Things (IoT), or Physical Internet (PI), are examined. A selection of "OpenAccess" publications was made from the scientific literature existing in some academic web portals. The selected publications were classified according to the following criteria: (prevalent) typology of content, PCS evolution (wave), port work-area, and geographical distribution of case studies [26].

The remaining part of the paper is articulated in four section. Section 2 describes the survey on the scientific literature about PCS and emerging ICT. Section 3 presents the distributions of the selected publication according to each selected criterion. The conclusions and the research perspective are reported in the last section.

2 SURVEY DESCRIPTION

The survey on the scientific literature about PCS and emerging ICT was conducted in two steps:

- Step I: search of scientific publications in the main academic web portals, according to specific keywords;

- Step II: classification of selected publications according to identified criteria.

Step I concerned the search of scientific publications, that included papers published in scientific journals, chapters of volumes, papers in conference proceedings.

The search was conducted through some keywords (or a combination of keywords) such as: "Port Community System", "Physical Internet" and "Port", and "Internet of Things" and "Port".

Preliminarily, one of the most popular academic web portals, "ResearchGate", was queried. It emerged that, by using the above keywords, the number of selected publications were very high. A more selective and in-depth search was performed on "Google Scholar", "Scopus" and "Web of Science" portals. Finally, a specific search was conducted in scientific journals belonging to the "Transportation" Q1-quartile of Scimago, finding very few publications.

As the keywords could recall publications that marginally deals with PCS and emerging ICT, papers without a specific focus on these issues were discarded, reducing the set of scientific publications to 170. This set was analysed and a further selection was made of publications that explicitly put in relation the PCS with port transport and logistics topics, selecting 53 publications. A final selection was made by introducing the criterion of "OpenAccess", obtaining a number of 43 scientific papers.

Step II concerned the classification of selected publications according to the following criteria.

a. *Contents*
   - Literature review (LT); state of the art on the topic of PCS and emerging ICT;
   - Theoretical (TH); theoretical approaches based on transport system model (TSMs);
   - Framework (FR); conceptual frameworks, or procedures, concerning the PCS, or concerning the use of emerging ICT within the PCS;
   - Case study (CS), case study of PCS and emerging ICT implemented inside ports;
   - Experimental data (ED), experimental data concerning the use of PCS, that allow a quantitative analysis of the impacts of PCS on port operations and on port stakeholders.

b. *PCS wave (emerging ICT)*

   The wave IV, according to Moros-Daza et al. [24], is characterized by the experimentation, and in cases the implementation, of PCS with the following emerging ICT categories: blockchain and artificial intelligence. The wave V, according to Russo and Musolino [22], is characterized by the initial experimentation (and in cases of implementation) of PCS with the enabling technologies already of wave IV, together with internet of things and big data technologies.

   According to the above considerations, the following two main classes are identified:

   - PCS-wave IV, if the publication deals with one of the following categories of emerging ICT (the relationship with PCS may be explicitly mentioned or not):
     - BlockChain (BC), to support for documental and financial transactions;
     - Artificial Intelligence (AI), to support forecasting and decision-making on maritime and land transport operations, as well as to automate some repetitive (albeit sensitive) manual activities, such as loading/unloading goods;
   - PCS-wave V, if the publication deals with one the following category of emerging ICT (the relationship with PCS may be explicitly mentioned, or not):

  - Internet of Things (IoT), to support the identification of "objects" present in ports and facilitate the exchange of information between them (IoT can be defined as a "transversal" technology, as it is related to the other emerging technologies);
  - Big data (BD), that enrich information from traditional data, in terms of volume and variety; and that are so quickly elapsed to improve the representation of real phenomena and their modelling.

A one more general class was identified, named PCS, which gather publications that deal with PCSs (potentially belonging also to I, II and III wave). However, it was not possible to associate the PCS described to any specific wave.

c. *Port work-area [21], [25]*
- Logistics and transport (L&T): logistics and transport of goods from land to sea (and vice versa) by road and rail transport (short and long distances);
- Customs (CU): import and export, administrative and custom control processes;
- Navigation (NA), management and planning of ship arrivals and departures;
- Dangerous goods (DG), traffic of dangerous goods, and related procedures.

d. *Geographical distribution of case studies:* North Europe (NE); Mediterranean (ME); Asia: Middle and far East (AS); North America (NA); South America (SA).

Table 1 shows the results of the classification of selected publications according to the above identified criteria.

3 ANALYSIS OF SELECTED PUBBLICATIONS

The section presents the distributions of the selected publication according to each selected criterion. It is worth noting that the selected publications generally belong to more than one identified category, as shown in Table 1. In this paragraph, however, a further filter has been introduced by identifying the prevalent type of contribution inside the publication.

3.1 Contents

The most numerous class is the CS one (44.4%), followed by FR (35.6%) one, as shown in Fig. 1. This is due to the fact that several studies are based on case studies of application of one single class of emerging ICT inside ports, followed by theoretical architectures (which can be explicitly related to a PCS in the publication, or not). The publications with theoretical content (TH) are a small percentage (5.8%), but not of lesser importance. The lack of theoretical publications is due to the fact that, both model specifications and experimental data, ED, (3.8%) related to PCS and emerging ICT are lacking. At the end, a number of publications, based on a literature review, are presents (11.1%).

3.2 PCS wave (emerging ICT)

The solution of stakeholders' interactions in port operations is a crucial point for increasing the overall port efficiency [21]. In this context, there are single categories of emerging ICT in the selected literature, which operate in relation with PCSs. There are no, or in any case no identifiable, ports that have internalized all emerging ICT inside a PCS.

The most common class of emerging ICT is IoT (39.5%), followed by BC (25.6%) and Big Data (20.9%), as showed in Fig. 2. IoT is currently one of the most developed emerging ICT, able to bring tangible benefits in port operations. BC is a technology that still does not

Table 1: Classification of selected publications according to the identified criteria.

Criteria	Publications				
	LT	TH	FR	CS	ED
Contents	[21] [23] [28] [29] [30] [32] [33] [34] [35] [38] [40] [44] [45] [46] [47] [48] [49] [51] [53] [54] [56] [57] [59] [63] [64] [67]	[21] [62] [68]	[23] [24] [27] [28] [29] [30] [31] [32] [33] [36] [37] [39] [40] [41] [42] [43] [44] [45] [46] [47] [49] [50] [52] [53] [55] [57] [58] [59] [60] [61] [63] [64] [66] [67]	[23] [27] [28] [30] [31] [33] [34] [35] [36] [37] [39] [40] [41] [43] [44] [45] [53] [54] [55] [57] [58] [59] [62] [68]	[38] [59]
	PCS	PCS-wave IV		PCS-wave V	
		BC	AI	IoT	BD
PCS wave (emerging ICT)	[24] [27] [28] [30] [32] [34] [35] [41] [46] [47] [49] [50] [52] [53] [54] [55] [56] [57] [58] [64] [65] [66]	[21] [36] [37] [44] [45] [51] [53] [54] [59] [62] [67]	[36] [37] [44] [48] [50] [51]	[21] [23] [29] [31] [33] [36] [37] [38] [40] [43] [50] [51] [53] [59] [61] [68]	[21] [23] [29] [33] [36] [37] [44] [61]
	L&T	CU	NA	DG	
Work-area	[27] [28] [29] [30] [31] [34] [35] [37] [39] [40] [41] [42] [43] [44] [45] [46] [47] [48] [49] [50] [51] [52] [53] [54] [57] [58] [59] [60] [61] [64] [65] [66] [67]	[34] [35] [41] [42] [44]	[27] [28] [42] [44]	[34] [41] [44] [34]	
	NE	ME	AS	NA	SA
Geographical distributions	[23] [31] [37] [40] [43] [48] [60]	[23] [27] [28] [30] [35] [37] [41] [42] [43] [50] [52] [53] [54] [58] [59] [64] [65]	[37] [53] [63]	[64]	[36] [55]

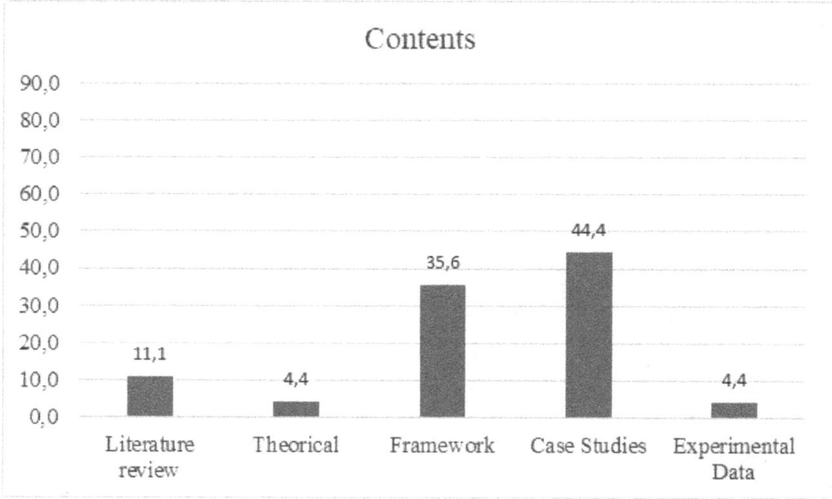

Figure 1: Distribution of selected publications for prevalent type of contents.

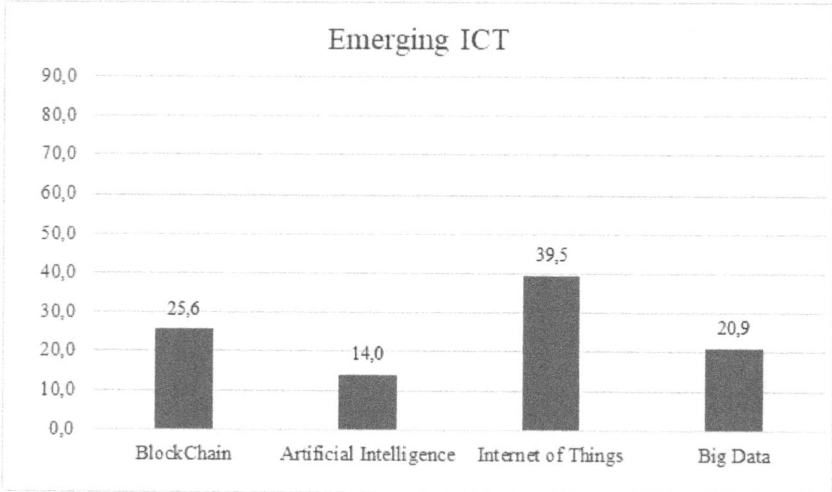

Figure 2: Distribution of selected publications for prevalent class of emerging ICT.

give sufficient guarantees from the point of view of security, and consequently no market applications are expected in the short term (e.g. bitcoin, smart contract). As far BD is concerned, the "cloud computing" represents today a valid solution, but it is not easy to implement, since it involves the information transfer and sharing by port stakeholders within the "cloud".

3.3 Port work-area

As shown in Fig. 3, the great part of the selected publications refers to the work-area "Logistics and transport" (84.8%), followed by the work-area "Custom" (9.1%).

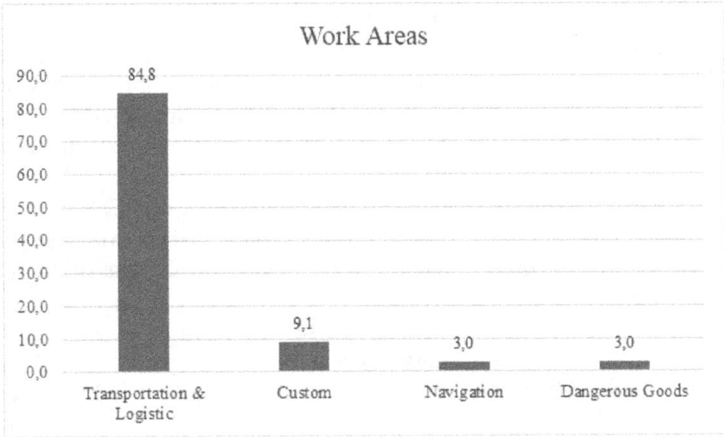

Figure 3: Distribution of selected publications for prevalent port work-area.

3.4 Geographical distribution

A further classification was made in relation to the geographical distribution of the case studies regarding the existing PCS. As showed in Fig. 4, many selected publications report case studies in European ports. The highest percentage concerns Mediterranean ports (40.0%), followed by Northern Europe ports (15.6%). These percentages are due to the fact that many European ports are already equipped by a PCS, as Rotterdam, Hamburg, Antwerp in Northern Europe. The reduced percentage of case studies related to Asian (6.7%) and North American (2.2%) may be misleading, but there is the awareness that both Asian ports (e.g. Shanghai, Singapore, etc.) and North American ones (e.g. Los Angeles, New York/New Jersey, etc.) are also equipped with PCS. There are case studies of PCS and emerging ICT related to South American ports (4.4%), indicating that also these ports are in line with the main ports.

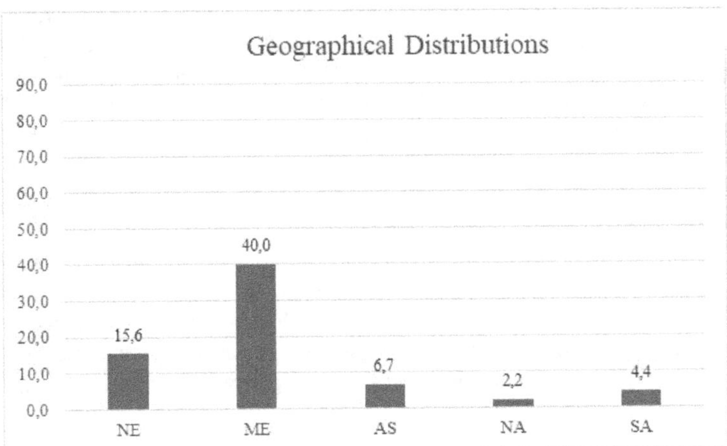

Figure 4: Distribution of selected publications according to geographical areas.

4 DISCUSSION AND CONCLUSIONS

Ports entered nowadays in a generation, the fifth, where benefits and challenges are connected with the introduction of emerging ICTs supporting interactions among port's stakeholders. The potentialities offered by emerging ICT are amplified by means of PCSs, which aim to increase cooperation between port stakeholders, and consequently to increase their competitiveness compared to other ports.

The paper presents a survey regarding the scientific literature on PCSs and the interactions between PCSs with emerging ICT. The survey was conducted in two steps: the search of scientific publications in the main academic web portals and the classification of selected publications according to identified criteria.

The main results of the survey are discussed in the following.

As far as concerns the analysis of single criterion, the "Contents" criterion shows that the greatest number of publications presents conceptual frameworks of PCSs, even if they are not prevalent in the publications. According to the criterion of "PCS wave", there are several publications dealing with PCS from which it is not possible clearly identify the wave. Among the emerging ICTs, the most numerous group of publications concerns the IoT. By considering the criterion of "Work area", the Logistics and transport is the one more studied among the selected publications. At the end, according to the "Geographical distribution" criterion, several publications deal with Mediterranean ports.

As far as concern analysis of combination of criteria, all couples of criteria were compared in order to determine the couple of classes (the "cells" in the Table 1), that have the highest number of shared publications. As result, the following main elements emerge. By comparing the couple of criteria "Contents-PCS wave (emerging ICT)" (excluding the more general category "PCS"), the couple "FR-IoT" presents the highest number of publications in common (12). The comparison of criteria "Contents-work area" identified the couple "FR-L&T" as the one with the highest number of publications in common (25). The comparison "PCS wave-work area" identified the couple "IoT-L&T" as the one with the highest number of publications in common (10). The comparison "PCS wave-geographical distribution" identified the couple "IoT-ME" as the one with the highest number of publications in common (7). The comparison "work area-geographical distribution" identified the couple "L&T-ME" as the one with the highest number of publications in common (16).

The above results show that there is an "ideal line connecting some emerging dots", along with the studies on PCS and emerging ICTs are concentrated. The "dots" are the conceptual frameworks (FR), the IoT technology (IoT), the Logistics and transportation work area (L&T), and the Mediterranean ports (ME). Therefore, the main conclusion is that there is a considerable interest in the combination of L&T with IoT, as core elements of advanced PCSs.

According to the above considerations, the future research could concern the quantitative analysis of the benefits of the IoT, and more generally of emerging ICTs, on L&T work area focusing on the interactions among the different stakeholders. The analysis could concern the quantitative estimation of the utility connected to goods handling and manipulation inside ports, in terms of added value growth and of reduction of port operations costs. The quantitative estimation could be supported by transportation system models (TSM) that incorporate the impacts of the emerging ICTs on port operations, on the stakeholders and on port functions.

REFERENCES

[1] UNCTAD, Port marketing and the challenge of the third generation port. Trade and Development Board Committee on Shipping ad hoc Intergovernment Group of Port Experts, 1994.

[2] UNCTAD, Fourth-generation Port: Technical note. Ports Newsletter No. 19, prepared by UNCTAD Secretariat, 1999.

[3] Russo, F. & G. Musolino, G., Quantitative characteristics for port generations: The Italian case study. *Int. J. TDI*, **4**(2), pp. 103–112, 2020.

[4] Davis, H.C., Regional port impact studies: A critique and suggested methodology. *Transportation Journal*, **23**, pp. 61–71, 1983.

[5] Mantecón, I.M., Millán, P.C., Castro, J.V. & González, M.Á.P., Economic impact of a port on the hinterland: Application to Santander's port. *IJSTL*, **4**(3), p. 235, 2012.

[6] Millán, P.C., Fuente, M.D.L., Fernández, X.L. & Pesquera, M.Á., The impact of deregulation in the hydrocarbon sector: Evidence at the main Spanish ports of import of hydrocarbons (1986–2013). *IJSTL*, **11**(1), p. 41, 2019.

[7] Zhong, H., Lin, Y. & Yip, T.L., A quantitative liability risk assessment of oil spills in oil ports: The case of the Guangdong–Hong Kong–Macao Greater Bay Area. *Maritime Policy and Management*, pp. 1–19, 2021.

[8] Russo, F. & Chilà, G., Structural factors for a third-generation port: Current state, limits and weaknesses of Gioia Tauro, Italy. *The Regional Transport Plan*, pp. 3–15, 2021.

[9] Russo, F. & Chilà, G., Structural factors for a third-generation port: Actions And measures for Gioia Tauro, Italy. *The Regional Transport Plan*, pp. 17–30, 2021.

[10] Russo, F., Panuccio, P. & Rindone, C., Structural factors for a third-generation port: Between hinterland regeneration and smart town in Gioia Tauro, Italy. *The Regional Transport Plan*, pp. 79–90, 2021.

[11] Russo, F. & Rindone, C., Structural factors for a third-generation port: Planning interventions for research and development in Gioia Tauro, Italy. *Ten-T Node*, pp. 67–78, 2021.

[12] Musolino, G. & Chilà, G., Structural factors for a third-generation port: Planning transport interventions in Gioia Tauro, Italy. pp. 31–42, 2021.

[13] Musolino, G. & Rosaria Trecozzi, M., Structural factors for a third-generation port: Planning interventions for agri-food logistics in Gioia Tauro, Italy. pp. 43–54, 2021.

[14] Musolino, G., Cartisano, A. & Fortugno, G., Structural factors for a third-generation port: Planning interventions for mechanical logistics in Gioia Tauro, Italy. pp. 55–66, 2021.

[15] Russo, F., Chilà, G. & Zito, C., Strategic planning for special economic zones to ports of the future: System of models and test case. *Computational Science and its Applications – ICCSA 2022 Workshops*, **13381**, pp. 173–184, 2022.

[16] Musolino, G., Cartisano, A. & Fortugno, G., Special economic zones planning for sustainable ports: Aggregate economic impact of port of Gioia Tauro. *Computational Science and its Applications – ICCSA 2022 Workshops*, **13381**, pp. 60–71, 2022.

[17] Musolino, D.A. & Panuccio, P., Special economic zones planning for sustainable ports: The test case of territorial attractiveness and urban planning in Calabria region. *Computational Science and its Applications – ICCSA 2022 Workshops*, **13381**, pp. 72–84, 2022.

[18] Rindone, C., Cirianni, F.M.M., Delfino, G. & Croce, A.I., Special economic zones planning for sustainable ports: The role of research and training in the Calabria region. *Computational Science and its Applications – ICCSA 2022 Workshops*, **13381**, pp. 85–97, 2022.
[19] Pellicanò, D.S. & Trecozzi, M.R., Special economic zones planning for sustainable ports: General approach for administrative simplifications and a test case. *Computational Science and its Applications – ICCSA 2022 Workshops*, **13381**, pp. 47–59, 2022.
[20] Russo, F. & Musolino, G., Case studies and theoretical approaches in port competition and cooperation. *Computational Science and its Applications – ICCSA 2021*, **12958**, pp. 198–212, 2021.
[21] Russo, F. & Musolino, G., The role of emerging ICT in the ports: Increasing utilities according to shared decisions. *Front. Future Transp.*, **2**, 722812, 2021.
[22] Russo, F. & Musolino, G., *Emerging ICT in Port Operations: Case Studies*, 1st ed., *Trends in Maritime Technology and Engineering Volume 2: Proceedings of the 6th International Conference on Maritime Technology and Engineering (MARTECH)*, 24–26 May, Lisbon, Portugal, CRC Press: London, 2022.
[23] Caldeirinha, V., Felício, J.A., Salvador, A.S., Nabais, J. & Pinho, T., The impact of port community systems (PCS) characteristics on performance. *Research in Transportation Economics*, **80**, 100818, 2020.
[24] Moros-Daza, A., Amaya-Mier, R. & Paternina-Arboleda, C., Port community systems: A structured literature review. *Transportation Research Part A: Policy and Practice*, **133**, pp. 27–46, 2020.
[25] Carlan, V., Sys, C. & Vanelslander, T., How port community systems can contribute to port competitiveness: Developing a cost–benefit framework. *Research in Transportation Business and Management*, **19**, pp. 51–64, 2016.
[26] Pedà Girolamo, *Porti di quinta generazione: analisi sperimentale dei tempi delle navi porta-container*, Tesi di Laurea, DIIES, Università Mediterranea di Reggio Calabria. Reggio Calabria (Italy), 2022.
[27] Barasti, D., Troscia, M., Lattuca, D., Tardo, A., Barsanti, I. & Pagano, P., An ICT prototyping framework for the "Port of the Future". *Sensors*, **22**, p. 246, 2022.
[28] Barbu, A., Militaru, G. & Deselnicu, D., An overview of the port community system from Romania. *The International Maritime Transport and Logistics Conference "Marlog 9" Impacts of the Fourth Industrial Revolution on Port-City Integration*, 2020.
[29] Bastug, S., Arabelen, G., Vural, C.A. & Deveci, D.A., A value chain analysis of a seaport from the perspective of Industry 4.0. *International Journal of Shipping and Transport Logistics*, **12**(4), pp. 367–397, 2020.
[30] Bisogno, M., Nota, G., Saccomanno, A. & Tommasetti, A., Improving the efficiency of port community systems through integrated information flows of logistic processes. *International Journal of Digital Accounting Research*, **15**, 2015.
[31] Bracke, V., Sebrechts, M., Moons, B., Hoebeke, J., De Turck, F. & Volckaert, B., Design and evaluation of a scalable Internet of Things backend for smart ports. *Software-Practice and Experience*, **51**(7), pp. 1557–1579, 2021.
[32] Caldeirinha, V., Nabais, J.L. & Pinto, C., Port community systems: Accelerating the transition of seaports toward the physical internet: The Portuguese case. *Journal of Marine Science and Engineering*, 2022.

[33] Chandra, D.R. & van Hillegersberg, J., Governance of inter-organizational systems: A longitudinal case study of Rotterdam's port community system. *International Journal of Information Systems and Project Management*, **6**(2), pp. 47–68, 2022.

[34] Di Vaio, A. & Varriale, L., AIS and reporting in the port community systems: An Italian case study in the landlord port model. *Lecture Notes in Information Systems and Organisation*, 2017.

[35] Di Vaio, A. & Varriale, L., Digitalization in the sea-land supply chain: Experiences from Italy in rethinking the port operations within inter-organizational relationships. *Production Planning and Control*, pp. 1–13, 2019.

[36] Durán, C., Córdova, F. & Palominos, F., A conceptual model for a cyber-social-technological-cognitive smart medium-size port. *Procedia Computer Science*, 2019.

[37] Elbert, R. & Tessmann, R., Port community systems: Supply chain app stores of the future? *International Logistics*, **73**, pp. 40–43, 2021.

[38] Fahim, P., Jayaraman, R., Poulin, M., Montreuil, B. & Tavasszy, L.A., The physical internet and maritime ports ready for the future. *IEEE Engineering Management Review*.

[39] Fahim, P., Rezaei, J., Montreuil, B. & Tavasszy, L., Port performance evaluation and selection in the physical internet. *Transport Policy*, 2021.

[40] Fahim, P.B.M., An, R., Rezaei, J., Pang, Y.S., Montreuil, B. & Tavasszy, L., An information architecture to enable track-and-trace capability in Physical Internet ports. *Computers in Industry*, 2021.

[41] Gabriel, R., Remus, Z., Acomi, N., Vasile, D. & Liviu, D., Prefeasibility study for port community system (PCS) in Constanta Port (Constanta, Midia, Mangalia). Danube Ports Network. (Program Transnational Duna rea 2014–2020)). Maritime Port Administration Constanta. Project co-funded by European Union funds (ERDF, IPA), 2018.

[42] Gattuso, D., Musolino, G. & Perri, V., Flussi informativi nel porto di Gioia Tauro. Progetto reports e progetto sestante. Azioni per lo sviluppo dello short sea shipping e per la sicurezza nei nodi portuali e per l'interoperabilità dei sistemi informativi. *Collana Trasporti*, ed. D. Gattuso, FrancoAngeli: Milan, pp. 167–184, 2005.

[43] Henríquez, F.R., Martínez de Osés, X. & Martínez Marín, J.E., Technological drivers of seaports' business model innovation: An exploratory case study on the port of Barcelona. *Research in Transportation Business and Management*, 100803, 2022.

[44] Henriquez, R., de Oses, F.X.M. & Marin, J.E.M., IoT-driven business model innovation a case-study on the port of Barcelona in the context of the belt and road initiative. Advances in E-Business Engineering For Ubiquitous Computing. *Lecture Notes on Data Engineering and Communications Technologies*, 2020.

[45] Irannezhad, E., Prato, C.G. & Hickman, M., An intelligent decision support system prototype for hinterland port logistics. *Decision Support Systems*, **130**, 113227, 2020.

[46] Irannezhad, E., The architectural design requirements of a blockchain-based port community system. *Logistics*, **4**, 2020.

[47] Jovic, M., Tijan, E., Br.cic, D. & Pucihar, A., Digitalization in maritime transport and seaports: Bibliometric, content and thematic analysis. *Journal of Marine Science and Engineering*, **10**, p. 486, 2022.

[48] Jović, M., Aksentijevic, S., Plentaj, B. & Tijan, E., Port community system business models. *34th Bled eConference Digital Support from Crisis to Progressive Change, Conference Proceedings*, pp. 41–52, 2021.

[49] Kaup, M., Deja, A., Slaczka, W. & Gróbarczyk, M., The port community system as an example of integration of port users. *Procedia Computer Science*, p. 192, 2021.

[50] Kenyon, G.N., Goldsmith, M., Neureuther, B.D., Zhou, D., Improving the return on investment in ports: Opportunities in data management. *Maritime Economics and Logistics*, 2018.

[51] Kumar, S. & Pundir, A., Application of IoT and blockchain technology in supply chain: A review. *GLOGIFT 19: Flexibility, Innovation and Sustainable Business*, 2019.

[52] Maeder, C., Sohr, K., Nguempnang, R.W., Meyer-Larsen, N. & Müller, R., Modeling and validating role-based authorization policies for a port communication system with UML and OCL. *Journal of Object Technology*, 2020.

[53] Mendes, C.J., Luncenti, K. & Deambrosi, S., International case studies and good practices for implementing port community systems, 2019.

[54] Meyer-Larsen, N. & Müller, R., Enhancing the cybersecurity of port community systems. *Dynamics in Logistics*, pp. 318–323, 2018.

[55] Moros-Daza, A., Amaya, R.A., Garcia, G. & Paternina, C., Assessing the effect of implementing a port community system platform in the response time of an international terminal: The case of a multi-cargo facility at the Colombian Caribbean coast. *Proceedings of the International Conference on Industrial Engineering and Operations Management*, 2016.

[56] Moros-Daza, A., Caicedo Solano, N., Amaya, R. & Paternina, C., A multivariate analysis for the creation of port community system approaches. *EURO Mini Conference on Advances in Freight Transportation and Logistics (emc-ftl-2018), Transportation Research Procedia*, **30**, pp. 127–136, 2018.

[57] Nadi, A., Sharma, S., Snelder, M., Bakri, T., van Lint, H. & Tavasszy, L., Short-term prediction of outbound truck traffic from the exchange of information in logistics hubs: A case study for the port of Rotterdam. *Transportation Research Part C: Emerging Technologies*, **127**, 103111, 2021.

[58] Nota, G., Bisogno, M. & Saccomanno, A., A service-oriented approach to modeling and performance analysis of port community systems. *International Journal of Engineering Business Management*, 2018.

[59] Pagano, P., Antonelli, S. & Tardo, A., C-Ports: A proposal for a comprehensive standardization and implementation plan of digital services offered by the port of the future. *Comput. Ind.*, **134**, 103556, 2021.

[60] Šakalys, A. & Greiciune, L., Innovative measures ensuring integrity of the entire supply chain or transport corridor. *Journal of Shipping and Ocean Engineering*, **5**, pp. 280–288, 2015.

[61] Sarabia-Jacome, D., Palau, C.E., Esteve, M. & Boronat, F., Seaport data space for improving logistic maritime operations. *IEEE Access*, **8**, 8946609, pp. 4372–4382, 2020.

[62] Schroten, A., Van Grinsven, A., Tol, E., Leestemaker, L., Schackmann, P.P., Vonk-Noordegraaf, D., Van Meijeren, J., Kalisvaart, S., Research for TRAN committee: The impact of emerging technologies on the transport system. European Parliament, Policy Department for Structural and Cohesion Policies: Brussels, 2020.

[63] Sunitiyoso, Y., Nuraeni, S., Pambudi, N., Inayati, T., Nurdayat, I., Hadiansyah, F. & Tiara, A., Port performance factors and their interactions: A systems thinking approach. *The Asian Journal of Shipping and Logistics*, 2022.

[64] Tijan, E., Jovic, M., Panjako, A. & Žgaljic, D., The role of port authority in port governance and port community system implementation. *Sustainability*, **13**, p. 2795, 2021.

[65] Torlak, I., Tijan, E., Aksentijevic, S. & Jugovic, A., Port community system feasibility analysis: Case study: Split. *43rd International Convention on Information, Communication and Electronic Technology (MIPRO)*, pp. 1410–1415, 2020.

[66] Tsamboulas, D., Moraiti, P. & Lekka, A.M., Performance evaluation for implementation of port community system. *Transportation Research Record*, 2012.

[67] Wang, J.-H. & Su, L.-H., Research on port supply chain logistics integration based on the internet of things, 2013.

[68] Yang, J., Zhang, X. & Yang, X., Evaluation of the development potential of Hainan Free Trade Port industrial chain driven by the Internet of Things distribution mechanism, 2022.

SECTION 4
MARITIME AND RAILWAYS TRANSPORTATION

SIGNIFICANCE AND CURRENT REGULATIONS OF EXTERNAL AIRBORNE NOISE FROM SHIPS

LUKA VUKIĆ, IVAN PERONJA & NIKOLA MANDIĆ
Faculty of Maritime Studies, University of Split, Croatia

ABSTRACT

Pollution sources of environmental noise are most extensive in the transport sector. Unlike other transport modalities, ship noise is given marginal importance in the scientific community and the public. External airborne ship noise is not yet categorized as an entity of environmental pollution but rather as a threat to passengers and crew in navigation and industrial noise in the port. Results of the research analysis indicate that official noise regulation in the maritime sector is modest, neglected, and outdated at the international and national levels. However, professional associations and agencies recognize the external airborne noise from ships as a significant source of environmental pollution that affects health, quality of life, and economy, respectively, a socio-economic component altogether. Croatia adopted a legal frame based on the Maritime Code that, although indirectly from the safety at work perspective, also regulates noise from ships. It is the only national, official document, especially concerning airborne ship noise as a significant source of pollution. A deficiency of the legal basis for measuring, locating, and identifying noise sources in this industry's sector implicates that vessels cannot be penalized according to the polluter pays principle, nor can ships with lower noise emissions be favored. It is necessary to close this regulation gap enabling the protection measures application following the sustainability principles in maritime transport.
Keywords: airborne noise, maritime traffic, regulations, ships.

1 INTRODUCTION

Maritime transport and port activities significantly contribute to environmental degradation and thus jeopardize the socio-economic benefits of transporting services and goods [1]. Noise as an output of ship activities affects the health of passengers and crew, increases background noise onshore and in ports, and negatively influences underwater life [2]. Since 1950, maritime background noise has doubled each decade due to the growth of marine traffic and ship capacities [3]. The actual noise emission level is highly dependent on the parameter of vessel type [4], so measurement of generated noise of each single pollution source is needed. At the same time, it is an obstacle to the acceptance of efficient and comprehensive regulations. Noise protection measures in maritime transport lag behind the ones in other transportation modalities. Recent scientific literature is relatively scarce, while at the same time, the volume of marine traffic increases continuously [5].

The ship at berth in port needs energy for electricity, ventilation, air conditioning, cooling, freezing, and others, usually supplied by an auxiliary engine, marked as primary noise sources. The noise power is proportional to the vessel's deadweight [6]. The auxiliary engine power can be up to 4 MW (for the largest cruisers) [7]. Some measures to reduce noise from ships comprise different types of mufflers on the exhaust pipe of the diesel generator and other machinery while in port. There is also an onshore power supply, ventilation damping, air conditioning systems, mineral wool coating in compartments where this machinery is located, barriers, louvers, etc. [8].

Tourism activities, freight transport, and passenger traffic in port cities are burdened with their own external airborne noise, disturbing domicile residents and generating night disturbances and sleep disorders [9]. Noise sources within a port area can be classified into subcategories: road, rail, ship, port, and industrial sources. This classification can be even

WIT Transactions on The Built Environment, Vol 212, © 2022 WIT Press
www.witpress.com, ISSN 1743-3509 (on-line)
doi:10.2495/UMT220121

more detailed based on the criteria of a specific vehicle or machine and its position within the port. Identifying the principal noise source for which complaints are filed simplifies the operation and enables targeted action [10]. Noise impact has recently become a priority as an environmental issue in seaports. Noise sources in the port are related to machinery operations, car and truck traffic, railway operations, ramps, cargo handling, and ships. The findings in the port of Stockholm, Helsinki, Tallinn, Turku, and Naantali, nominated as PENTA ports, indicated the RoPax operations as the largest identified noise source. From the perspective of noise in the port area, there is an unambiguous connection between noise pollution and urban planning [11].

Unlike other forms of pollution, the impact of noise depends not only on the physical characteristics of the sound but even more on the human reaction to it. Impacts on human health and quality of daily living are those items against which the impact of noise is measured, and to a significant extent, they are burdened with subjectivity. Some people are more disturbed by occasional noise, while others are irritated by persistent background noise. There are also differences in experiences with low and high-frequencies sounds. In overcoming subjectivity, the effects of the highest sound level should always be measured, thus eliminating noise that causes annoys [12]. Exposure to noise larger than 85 dB and more than 20 years is associated with a risk of arterial hypertension [13]. Although this claim has never been proven in the maritime sector [14], measures taken on noise pollution in marine transport should respect this cause-and-effect relationship. In a study on offshore supplying vessels, hearing impairments in the low and high frequencies scales were identified as a possible consequence of prolonged exposure and high noise levels [15]. In maritime traffic, the crew is constantly exposed to vibration, noise, and heat. When examining these factors individually onto their impact on health and quality of life, the context of simultaneous action of several factors needs to be a framework for discussion. A short-term study [16] already revealed a substantial level of stress.

For the International Maritime Organization (IMO), airborne noise pollution from ships is primarily a matter of occupational health. When the ship enters the port, port noise is classified as industrial noise. Noise emissions from maritime transport are ordinarily regulated at the national level. Due to all this, noise has not a high priority for ship-owners [11]. There is no regulation of airborne noise emission for every individual ship type. There is an initiative to mark them due to the emitted noise as Super Quiet, Quiet, Standard, Inland waterways, and Commercial [17]. Some measures to reduce port noise are convenient port location, regulation of port traffic, ramp design, speed reduction, modernization of machinery, walls, noise barriers, and measures on berths such as distance, slot use, working time restrictions, etc. Cruise and RoPax ships are paid special attention as they berth in the places closest to the city [18].

2 REGULATIONS

There is no universal protocol for measuring airborne noise emissions from ships and standardized data regarding noise and nuisance in and around the port to date. Emissions are regulated at the national level or through port environmental permits. They differ in different states and even in ports of the same ones. Airborne noise emissions are not currently on the IMO agendas, and no regulation should be expected soon [19]. Project NEPTUNE aims to classify the emitted noise from different ships according to the total noise power and the share of low-frequency noise. It also seeks to incorporate this data into the Environmental Ship Index (ESI). The first ship was already certified in September 2020 [20], [21]. For a complete picture of ambient noise, it is necessary to perform numerous measurements at many locations for at least one year [22].

Current regulations regulate exposure to noise of crew and passengers on board, but not workers and residents in port cities. The problem is exacerbated by the lack of jurisdiction over foreign-flagged ships in free zones [23], vessels used for non-commercial purposes, and in cases of innocent passage through the territorial sea [24]. Installing shore power is not only expensive, but most of all, it radically reduces only air pollution and not noise in the same way [25]. For now, it only makes sense on Ro-Ro terminals. ESI does not envisage noise limits nor a Green Award System, based on which port dues for green boats could be reduced and thus stimulate the installation of silencers [6]. Although the port is considered an "industrial plant" according to the current regulations [17], Lloyd Register noticed the significance of external airborne noise emission at the port area within the total environmental noise [26]. According to noise being a transboundary polluter, it should be regulated internationally. The ports are responsible for noise pollution but cannot affect it precisely because of the lack of international regulations [27]. The administrative framework of rules and regulations on port noise is shown in Fig. 1.

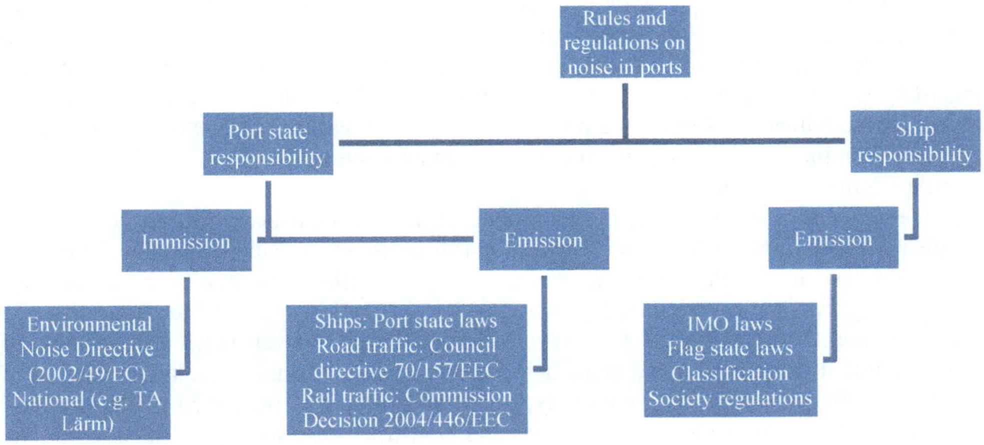

Figure 1: Rules and regulations on port noise [28].

Global cooperation on this issue is difficult to achieve. There are no international standards for port noise management. The port noise responsibility is mainly allocated to the ports and maritime industry, while the IMO is responsible for noise regulation on board. Only a few ports have implemented measures to reduce noise. Rules vary and are applied differently [29].

2.1 Regulations at the European level

The fundamental document that regulates noise on ships is IMO Resolution MSC.337 (91), "Code on noise levels on board ships" (adopted on 30 November 2012), which is in force from 1 July 2014 and is mandatory for new vessels. It is compulsory based on the International Convention for the Safety of Life at Sea (SOLAS) II-1/3-12 [30]. In Annex 1 of the IMO, Circular no. 3644 of 20 May 2016, amendments to Rule II-1/3-12 were made [31].

For ships in use before, in some segments, Resolution A.468 (XII) since 19 November 1981 has still been in force [32]. In Directive 2002/49/EC, the Commission considers noise pollution to be one of the biggest environmental problems in Europe, highlighting transport activity as one of the principal sources of environmental noise. The Directive explicitly lists all transport modes, but not maritime transport, nor even within industrial noise context [33].

At the International Labour Conference in 2006, a valuable Maritime Labour Convention was adopted, which in significant segment related to protection against noise and vibration on ships [34]. Noise is also well addressed in the European Standard laying down Technical Requirements for Inland Navigation vessels [35] but does not apply to maritime transport. In the 2018 report, the European Environment Agency (EEA) [36] emphasizes environmental pollution from the air and marine emissions, including noise pollution.

2.2 Regulations at the national level

In the Republic of Croatia, noise is regulated by the Noise Protection Act [37] and the following Ordinances: the Ordinance on maximum permissible noise levels in the environment in which people work and reside, the Ordinance on the protection of workers from noise exposure at work, the Ordinance on activities for which it is necessary to determine the implementation of noise protection measures, the Ordinance on the conditions regarding space, equipment and employees of legal entities performing professional noise protection activities, the Ordinance on noise protection measures of open-sources and the Ordinance on the preparation and content of noise maps and action plans and the method of calculating the permissible noise indicators [38].

Amendments to the Act [39] have brought alignment with the European Union (EU) acquis in this area, and no document mentions maritime transport. Although in the elementary text of the law, in Article 2, item 1, marine and river traffic is listed as sources that emit environmental noise, in Article 1, the same law states that this law does not apply to noise caused by maritime facilities [37]. In Articles 18 and 21 of the Ordinance on the method of production and content of noise maps and action plans and on the calculation method of permitted noise indicators, unlike other types of transport, maritime and river transport and ports are explicitly linked to industrial areas and industrial noise sources [40]. Based on the Maritime Code [41], the Ministry of the Sea, Transport, and Infrastructure has adopted Rules for the Statutory Certification of Seagoing Ships with the highest permissible noise assessed levels and within the scope of occupational safety and crew accommodation [42]. It is the only document that regulates external airborne noise from ships specifically. The Rules for Statutory Certification of Passenger Ships in National Navigation in 2017 [43] contain elementary articles on noise protection, while the current version [44] refers to EU directives and repeals specific paragraphs. In other official legal regulations of the Republic of Croatia, noise from ships has a marginal meaning.

3 DISCUSSION

Noise is recognized as environmental pollution, which sources are most extensive and most common in the transport sector. Unlike other transport modalities, noise from ships is given marginal importance in the scientific community and the public. Except for the underwater noise, external airborne ship noise is not yet accepted as an entity of environmental pollution but as a threat to passengers and crew in navigation mode and as industrial noise in the port. Based on this noise hazards understanding, legal regulations have been adopted at the European and national levels. Results of the analysis indicate that noise regulation in the maritime sector is modest, neglected, and outdated. It can be noticed that the Ordinances of administrative bodies at the European level date from 1981, 2002, and 2012 and are still in force. External airborne noise from ships is hardly mentioned in recent regulations. However, international associations and agencies recognize the external airborne noise from ships as a significant source of environmental pollution that affects health, quality of life and economy, or socio-economic components altogether. At the national level, and especially after the acceptance and alignment with the legal regulations of the European Union, environmental pollution by noise from ships is marginalized in the Law on Noise Protection. External airborne noise from vessels is recognized in some segments of this law but is not regulated. In 2018, the Ministry of the Sea, Transport, and Infrastructure, based on the Maritime Code, adopted Rules that, although from the position of safety at work, also regulate noise from ships. In Croatia, documentation from a governing body addresses the shortcomings which the documents of non-governmental organizations and agencies fill at the European level. As long as external airborne noise from ships is not considered an environmental problem, positive measures cannot perform. Vessels cannot be sanctioned according to the polluter pays principle, nor can ships with lower noise emissions be rewarded. It is a deficiency implication of the legal basis for measuring, locating, and identifying noise sources in this segment. Sustainable development policy in the maritime transport industry dictates such a need.

There is no single standard for measuring ship noise, nor are the critical activities such as maintenance and loading and unloading included in such measurement. The night limit in cities does not exceed the value of the annual noise average of 40 dB. Above this sound power limit, sleep disturbances and annoyance can occur, and it can be assumed that this noise originates from ships in city ports as a primary source [45]. The sound power of the generator exhaust or fan wings can exceed 100 dB. In those circumstances, the calculation determined that such a ship should be moored at least 600 m from the settlement [8]. Noise maps of port terminals are a good tool for planning and predicting the impact of noise on the city and its surroundings [46].

Noise as a polluter seems underestimated in public compared to CO_2 emissions and climate change topics. The analysis shows that the implications of noise pollution in maritime transport observed ten years ago are still relevant today in the same form. Subsequent installation of silencers is quite expensive, resulting in a reduction in noise of only a few dB. Given the lifespan of commercial ships of close to 30 years, the fleet renewal will not be rapid. There is also a lack of awareness among citizens and administration and measures to reduce noise [5]. Pettersen [4] believes a good balance between efficiency and low noise can be an engineering challenge. Apart from technical improvements and advances in regulations and awareness-raising, the stakeholders' scale priorities and the scientific community is modest. Discounts on port dues and shorter berthing and dwell times are several incentives to reduce ship noise. Canadian ports offer attractive port dues for vessels that emit lower underwater noise [4] but not airborne noise emissions. Cooperation and joint action of ports are also expected to implement collective measures through environmental indices and social regulations. Best practices and experience acquired with airborne noise near airports should be exploited [47]. Noise maps help locate remediation priorities. Better cooperation between ports, shipping companies, and the city is needed, and noise management should become a part of socially responsible work. Noise management is more of an opportunity and less of a threat to the port industry [11]. In the recent report, the EEA considers that not just one but a combination of noise abatement measures is sufficient, which includes technological improvements, ambitious noise policy, better urban and infrastructural planning, and changes in human behavior [48]. EU policy for shifting road transport towards more environmentally friendly transport modes, i.e., maritime and rail transport, seems contradictory with the sustainable development guidelines of ports and marine transport as for the increased ports and seas pollution, including noise pollution [27]. The World Health Organization (WHO) rates noise as a significant public health problem of growing concern. Recently, updated guidelines have been adopted, including new noise sources, but maritime transport is omitted again [49]. The reasons could be in comparative research with other noise sources. Industrial noise, which includes port noise, minimally contributes to harmful noise exposure as it affects only 0.15% of the urban population per day and 0.08% per night [48]. In the chronological analysis of the legal regulations of transport noise in England from medieval times until today, nothing was found relating to ships noise [50]. For now, underwater noise remains the only satisfactorily regulated segment of noise pollution in maritime transport. Despite the facts mentioned above, the American Bureau of Shipping issued a guide that equates the importance and threat of underwater and external airborne noise from ships [51]. Better education and awareness-raising on the harmfulness of noise are likely to be crucial for changing attitudes towards noise pollution in maritime transport [52].

4 CONCLUSION

Noise does not affect only health but the entire environment. It reduces the quality of life and endangers some segments of the economy. Classified as industrial noise, external airborne noise from ships in ports is only one of its components. But really, it presents an environmental noise structural element not regulated after its meaning. Non-governmental organizations and insurance companies have already recognized the problem. Ports responsible for noise pollution cannot be sanctioned or modernized due to the lack of international regulations. European regulations are flawed and outdated, dating from 2014 and 2016 for new ships and from 1981 for older ones. The only document that specially regulates external airborne ship noise at the national level is the Ordinance in the framework of occupational safety. Exposure to the noise of the crew and passengers on the ship on the one hand and workers and residents in port cities on the other is not legally equalized. The ESI and Green Awards System projects have no operational but only nominal meaning. It is necessary to close this gap enabling the protection measures application following the sustainability principles in maritime transport. The measurement of airborne noise emission from ships needs to be standardized, and the regulations at the international and national level and among ports updated and harmonized.

ACKNOWLEDGEMENT

This work was supported by the funding from the University of Split, Faculty of Maritime Studies on the scientific project "Measurement and validation of environmental noise in the Port of Split"; Class; 641-02/21-01/0001; Reg. No.: 2181-197-01-06-21-0003.

REFERENCES

[1] Boteler, B., Grünig, M., Lago, M., Campos, A.I., Reker, J. & Meiner, A., European maritime transport and port activities: Identifying policy gaps towards reducing environmental impacts of socio-economic activities. Benjamin Boteler Ecologic Institute, Berlin, Germany, 2012. https://www.ecologic.eu/sites/default/files/presentation/2014/european-maritime-transport-and-port-activities_0.pdf. Accessed on: 25 May 2022.

[2] Badino, A., Borelli, D., Gaggero, T., Rizzuto, E. & Schenone, C., Noise emitted from ships: Impact inside and outside the vessels. *Procedia – Social and Behavioral Sciences*, **48**, pp. 868–879, 2012. DOI: 10.1016/j.sbspro.2012.06.1064.

[3] Jones, N., Ocean uproar: Saving marine life from a barrage of noise. *Nature*, **568**, pp. 158–161, 2019. DOI: 10.1038/d41586-019-01098-6.

[4] Pettersen, Ø.S., First DNV SILENT-E class notation awarded to a merchant vessel DNV. Marine impact. https://www.dnv.com/expert-story/maritime-impact/First-DNV-SILENT-E-class-notation-awarded-to-a-merchant-vessel.html. Accessed on: 1 Jun. 2022.

[5] Bolognese, M., Fidecaro, F., Palazzuoli, D. & Gaetano Licitra, G., Port noise and complaints in the North Tyrrhenian Sea and framework for remediation. *Environments*, **7**(2), p. 17, 2020. DOI: 10.3390/environments7020017.

[6] Witte, R., Regulation of noise from moored ships in ports. *EuroNoise 2015*, 31 May–3 Jun., Maastricht, 2015. https://www.conforg.fr/euronoise2015/proceedings/data/articles/000311.pdf. Accessed on: 2 Jun. 2022.

[7] Jalkanen, J.P., Brink, A., Kalli, J., Pettersson, H., Kukkonen, J. & Stipa, T., A modelling system for the exhaust emissions of marine traffic and its application in the Baltic Sea area. *Atmospheric Chemistry and Physics*, **9**, pp. 9209–9223, 2009. DOI: 10.5194/acp-9-9209-2009.

[8] Danish Ministry of the Environment, Noise from ships in ports: Possibilities for noise reduction. Environmental Project No. 1330, Lloyd's Register ODS, Denmark, 2010. https://mst.dk/media/mst/66165/978-87-92668-35-6.pdf. Accessed on: 3 Apr. 2022.

[9] Schenone, C., Borelli, D., Pallavidino, E. & Yousseu, A., The port noise analysis and control in Interreg Italy–France maritime programme, noise control for a better environment. *Inter-Noise 2019*, 16–19 Jun., Madrid, Spain, 2019. http://www.sea-acustica.es/fileadmin/INTERNOISE_2019/Fchrs/Proceedings/1898.pdf. Accessed on: 3 Apr. 2022.

[10] Fredianelli, L., Bolognese, M., Fidecaro, F. & Licitra, G., Classification of noise sources for port area noise mapping. *Environments*, **8**(2), p. 12, 2021. DOI: 10.3390/environments8020012.

[11] Mustonen, M., Noise as an environmental challenge for ports. TFK Rapport 2013:X, TFK Transport Research Institute, Stockholm, Sweden, 2013. https://projects.centralbaltic.eu/images/files/result_pdf/PENTA_result4_noise.pdf. Accessed on: 5 Apr. 2022.

[12] UK Civil Aviation Authority, Measuring and modelling noise: How aviation noise can be measured and modeled. https://www.caa.co.uk/Consumers/Environment/Noise/Measuring-and-modelling-noise/. Accessed on: 15 Apr. 2022.

[13] Jégaden, D., Concerning the relationship between noise on board ship and the onset of hypertension in seafarers. *Archives of Preventive Medicine*, **6**(1), pp. 001–002, 2021. DOI: 10.17352/apm.000026.

[14] Nguyen, T.P.N. et al., Noise exposure and its relationship with hypertension among fishermen in Thua Thien Hue Province, Vietnam. *Journal of Integrated Community Health*, **9**(1), pp. 3–16, 2020. DOI: 10.24321/2319.9113.202001.

[15] Liu, H., Yan, M.F., Ning, Y., Hou, X., Hou, B. & Tang, N.J., Noise exposure and hearing conditions of crew on offshore support vessels for petroleum production. *Journal of Environmental and Occupational Medicine*, **36**(5), pp. 474–478, 2019.

[16] Oldenburg, M., Felten, C., Hedtmann, J. & Jensen, H.J., Physical influences on seafarers are different during their voyage episodes of port stay, river passage and sea passage: A maritime field study. *PLoS One*, **15**(4), e0231309, 2020. DOI: 10.1371/journal.pone.0231309.

[17] The Maritime Executive, Industry-first airborne noise notation for ships in port launched. https://maritime-executive.com/article/industry-first-airborne-noise-notation-for-ships-in-port-launched. Accessed on: 5 Mar. 2022.
[18] Marrero, A., Ports seek to reduce acoustic pollution. PierNext 2020, 2020. https://piernext.portdebarcelona.cat/en/environment/ports-seek-to-reduce-acoustic-pollution/. Accessed on: 10 Mar. 2022.
[19] Bergh, Å., Muffling the rumble. Maritime impact, DNV AS., 2019. https://www.dnv.com/expert-story/maritime-impact/QUIET-A-DNV-GL-class-notation-for-airborne-external-noise-from-ships.html. Accessed on: 10 May 2022.
[20] World Port Sustainability Program (WPSP), Defining, measuring and mitigating vessel noise at ports. https://sustainableworldports.org/defining-measuring-and-mitigating-vessel-noise-at-ports/. Accessed on: 11 May 2022.
[21] NEPTUNES, First vessel noise certified by ESI. https://neptunes.pro/2021/02/09/first-vessel-noise-certified-by-esi/. Accessed on: 20 May 2022.
[22] Harland, E.J., Jones, S.A.S. & Clarke, T., SEA 6 Technical report: underwater ambient noise. Farnborough, UK, Qinetiq, 2005. https://assets.publishing.service.gov.uk/government/uploads/system/uploads/attachment_data/file/197303/SEA6_Noise_QinetiQ.pdf. Accessed on: 20 May 2022.
[23] Omi, K., Extraterritoriality of free zones: The necessity for enhanced customs involvement. WCO Research Paper No. 47, 2019. http://www.wcoomd.org/-/media/wco/public/global/pdf/topics/research/research-paper-series/47_free_zones_customs_involvement_omi_en.pdf?la=en. Accessed on: 1 Feb. 2022.
[24] NOAA Office of General Counsel, Jurisdiction over vessels. https://www.gc.noaa.gov/gcil_jurisdiction.html. Accessed on: 3 Mar. 2022.
[25] Witte, R.J., Noise from moored ships. *InterNoise 2010*, 13–16 Jun., Lisbon, Portugal, 2010. https://dgmr.nl/wp-content/uploads/files/Internoise_2010_Rob_Witte_Noise_from_moored_ships.pdf. Accessed on: 5 Mar. 2022.
[26] Lloyd's Register, LR launches industry first airborne noise notation for ships in ports. https://www.lr.org/en/latest-news/first-airborne-noise-notation-for-ships-in-ports/. Accessed on: 5 Mar. 2022.
[27] Giannoumis, J.K., Underwater noise pollution: An evaluation of the voluntary guidelines on reduction of underwater noise from commercial shipping in the United States. Master's thesis, Noragric Ås, Akershus, Norwegian University of Life Sciences, Norway, 2017.
[28] Green Cruise Port, Hamburg Port Authority and DW-Ship-Consult GmbH: Emission sources and possible mitigation measures of cruise terminals. http://greencruiseport.eu/files/public/download/studies/HPA%20GCP%20report%20Part%20A_Emission%20Sources%20and%20Possible%20Mitigation%20Measures%20of%20Cruise%20Terminals_Final2_26.04.2018.pdf. Accessed on: 10 Jun. 2022.
[29] Shi, Q. & Xu, S., From a global review of port noise management initiatives to a specific port city's case study. MSc thesis, School of Business, Economics and Law, University of Gothenburg, Sweden, 2019. https://gupea.ub.gu.se/bitstream/handle/2077/60878/gupea_2077_60878_1.pdf?sequence=1. Accessed on: 12 Jun. 2022.
[30] International Maritime Organization (IMO), Adoption of the code on noise levels on board ships. https://wwwcdn.imo.org/localresources/en/KnowledgeCentre/IndexofIMOResolutions/Documents/MSC%20-%20Maritime%20Safety/337(91).pdf. Accessed on: 15 Jun. 2022.

[31] European Commission, COM (2016) 549 final, 2016/0263 (NLE), 30.8.2016 Brussels. https://eur-lex.europa.eu/legal-content/EN/ALL/?uri=COM:2016:549:FIN. Accessed on: 15 Jun. 2022.

[32] The Netherlands Human Environment and Transport Inspectorate, Overview noise level requirements. https://english.ilent.nlpublications2016/04/19. Accessed on: 15 Mar. 2022.

[33] European Parliament, EU directive 2002/49/ec of the European parliament and of the Council of 25 June 2002 relating to the assessment and management of environmental noise. Official Journal of the European Communities L 189/12-25, 2002. https://eur-lex.europa.eu/LexUriServ/LexUriServ.do?uri=OJ:L:2002:189:0012:0025:EN:PDF. Accessed on: 20 Mar. 2022.

[34] International Labour Conference, Maritime Labour Convention. http://www.ilo.org/wcmsp5/groups/public/%40ed_norm/%40normes/documents/normativeinstrument/wcms_090250.pdf. Accessed on: 16 Mar. 2022.

[35] European Committee for Drawing up Standards in the field of Inland Navigation (CESNI), European standard laying down technical requirements for inland navigation vessels (ES-TRIN). https://www.cesni.eu/wp-content/uploads/2020/10/ES_TRIN_2021_en.pdf. Accessed on: 16 Mar. 2022.

[36] European Environment Agency (EEA), Aviation and shipping emissions in focus. Copenhagen, Denmark. https://www.eea.europa.eu/articles/aviation-and-shipping-emissions-in-focus. Accessed on: 16 Mar. 2022.

[37] Croatian Parliament, Noise Protection Act. Official Gazette 30/09, 55/13, 153/13, 41/16, 114/18, 14/21, Zagreb, Croatia, 2009.

[38] Croatian Chamber of Architects, Laws and Regulations, Noise protection. https://www.arhitekti-hka.hr/hr/zakoni-propisi/popis/zastita-od-buke. Accessed on: 17 Mar. 2022.

[39] Croatian Parliament, Amendments to Noise Protection Act. Official Gazette 14/2021-272, Zagreb, Croatia, 2021.

[40] Minister of Health and Social Welfare of the Republic of Croatia, Ordinance on the method of production and content of noise maps and action plans and on the method of calculation of permitted noise indicators, Editorial consolidated text. Official Gazette 75/09, 60/16, 117/18, Zagreb, Croatia, 2009.

[41] Croatian Parliament, Maritime Code. Official Gazette 181/04, 76/07, 146/08, 61/11, 56/13, 26/15, 17/19, Zagreb, Croatia, 2004.

[42] Ministry of the Sea, Traffic and Infrastructure of the Republic of Croatia, Rules for statutory certification of seagoing vessels, safety at work and crew accommodation. Official Gazette 32/2018-650, Zagreb, Croatia, 2018.

[43] Ministry of the Sea, Traffic and Infrastructure of the Republic of Croatia, Rules for statutory certification of passenger ships in national navigation. Official Gazette 73/2017-1778, Zagreb, Croatia, 2017.

[44] Ministry of the Sea, Traffic and Infrastructure of the Republic of Croatia, Rules for statutory certification of passenger ships in national navigation. Official Gazette 106/2021-1863, Zagreb, Croatia, 2017.

[45] Marcellin, F., NEPTUNES: Measuring disruptive noise pollution from ships. Ship Technology, 2019. https://www.ship-technology.com/features/noise-pollution-from-ships/. Accessed on: 22 Mar. 2022.

[46] I-Hung, K. & Tang-Hung, N., Development and validation of noise maps for the container terminals at the port of Long Beach. *International Journal of Environmental Pollution and Remediation (IJEPR)*, **2**, pp. 1–12, 2014. DOI: 10.11159/ijepr.2014.001.

[47] Virto, L.R., Dumez, H., Romero, C. & Bailly, D., How can ports act to reduce underwater noise from shipping? Identifying effective management frameworks. Institut interdisciplinaire de l'innovation, Working Paper 21-CRG-01, FAERE – French Association of Environmental and Resource Economists, 2021. https://i3.cnrs.fr/wp-content/uploads/2021/05/Ports-can-reduce-underwater-L.-Recuero-Virto-H.-Dumez-C.-Romero-D.-Bailly.pdf. Accessed on: 21 Mar. 2022.
[48] Peris, E. (ed.), Environmental Noise in Europe-2020. EEA Report No 22/2019, Copenhagen, Denmark, 2020. https://www.eea.europa.eu/publications/environmental-noise-in-europe. Accessed on: 23 Mar. 2022.
[49] World Health Organization (WHO), Environmental Noise Guidelines for the European Region. Copenhagen, Denmark, 2018. https://www.euro.who.int/en/publications/abstracts/environmental-noise-guidelines-for-the-european-region-2018. Accessed on: 23 Mar. 2022.
[50] Simona, J., Rossell, J.M., Sanchez-Roemmele, X. & Vallbé, M., Evolution in the law of transport noise in England. *Transportation Research Part D: Transport and Environment*, **100**, 103050, 2021. DOI: 10.1016/j.trd.2021.103050.
[51] American Bureau of Shipping, ABS guide for the classification notation underwater noise and external airborne noise. https://ww2.eagle.org/content/dam/eagle/rules-and-guides/current/other/295-guide-classification-notation-underwater-noise-and-external-airborne-noise-2021/uwn-airn-guide-may21.pd. Accessed on: 23 May 2022.
[52] Vukić, L., Mihanović, V., Luca Fredianelli, L. & Plazibat, V., Seafarers' perception and attitudes towards noise emission on board ships. *International Journal of Environmental Research and Public Health*, **18**(12), p. 6671, 2021. DOI: 10.3390/ijerph18126671.

DEVELOPING ACCOUNTABLE MARITIME TRANSPORT AND PORT ORGANIZATIONAL STRUCTURES IN ARAB COUNTRIES

KHALED EL SAKTY[1] & ALIA EMAD ISLAM[2]
[1]College of International Transport and Logistics,
Arab Academy for Science, Technology and Maritime Transport, Egypt
[2]Department of Logistics of International Trade Management, College of International Transport and Logistics,
Arab Academy for Science, Technology and Maritime Transport, Egypt

ABSTRACT

The absence of specific goals, functions, integrity and a clear chain of command in the maritime transport system organizational structure (OS) in the Arab world may lead to creating trade tensions, increasing export restrictions, increasing economic and social costs, and other obstacles that impede the achievement of the strategic goal of maritime transport. This research aims to propose an integrated organizational structure that shows the relations and dependencies of all elements in the maritime transport system and the ports authorities, which helps in creating a stronger internal organizational environment. The proposed OSs are meant to improve maritime activities such as shipbuilding and ship scrapping, and to increase coordination and integration between the maritime transport sector and the authority for maritime safety through identifying the role of each entity within the proposed OSs. It is an exploratory multiple case study research, where a qualitative research method is conducted using interviews. The research approach has taken the following steps: (1) 47 world maritime transport organizational structures (MTOSs) that are selected and criticized according to six criteria; (2) data and information of 47 MTOSs were collected using documentation from official websites of governments; (3) the characteristics and challenges facing current MTOSs are identified; (4) the proposed MTOSs are projected; (5) purposive expert sampling technique is used for the validation of the proposed MTOSs and POS through the conduction of semi structured and unstructured interviews; (6) the proposed model is readjusted and implemented in Egypt according to the experts' feedback; and (7) external and internal validity were also tested to ensure that generalizability of the research is shown within its scope.

Keywords: accountability, governance, government performance, new public management, organizational structure, reengineering and restructuring.

1 INTRODUCTION

One of the problems facing organizations is the lack of a well-defined organizational structure that helps increases performance and productivity as well as delivering a valuable and quality service/product. The field of transportation and the maritime sectors in governments are not different from this concept. An integrated organizational structure helps in setting a healthier internal organizational environment for the sector and consequently enhancing measuring and managing performance [1]. Performance management is important for governments as it helps in presenting and addressing the issues of accountability and transparency [2], [3], while the more control a government has over public servants, the greater accountability it has [4]. New public management and accountability are strongly related [5], [6]; in addition, accountability is the possibility of holding the public servants responsible for their actions [7].

The government's initial aim is to reach public accountability that may be achieved through measuring and managing performance [6]. This is emphasized even more through 14 key elements that are essential for governmental performance measurement and management [8]. One of those elements is about precisely knowing the responsibilities of

WIT Transactions on The Built Environment, Vol 212, © 2022 WIT Press
www.witpress.com, ISSN 1743-3509 (on-line)
doi:10.2495/UMT220131

each person in the government to identify rewards and punishments that consequently build accountability in a country, which helps in maintaining control of public servants [4]. Consequently, this highlights the importance of having a clear organizational structure (OS) for all governmental bodies, departments, and sectors, provided that this OS is characterized by the ability to provide governance.

Accordingly, the governmental OS and planning became important to maintain political, social, and economic stability within a country. Therefore, one of the governmental administrative reform challenges is the lack of governmental OS that can apply the principles of governance and accountability. According to Islam [9], this can be due to the following three main problems: (i) the existence of the "many hands" problem in the various governmental structures, which prevents knowing the responsibilities of each individual in the government. This leads to the inability to fully define the rewards and penalties; (ii) the frequent modification of governmental structures through the addition, transfer, or deletion of different departments, which is followed by amendments and restructuring in the various departments and units. This weakens government construction and reduces the stability of employees and thus affects government performance in general; and (iii) The lack of connection and integration among the different departments and organizational divisions within the same organizational structure, or among the different organizational structures within the same governmental authority, which results in the waste of the state's economic and social resources.

Among different governmental OSs, maritime transport is considered as one of the most important modes of transport in transporting raw materials, commodities, bulk, packed and containerized goods to meet the needs of operating countries and companies. Maritime transport was responsible for carrying out a massive volume that reached nearly 11.8 billion tons in 2019 and carried out about 811.2 million TEUs that were traded for the same year [10]. In addition, ports represent an important element in the maritime transport industry because of their economic and operational roles that are reflected in the cost and time of loading and unloading of goods. This affects the performance of markets and meeting the needs of countries and companies from the timing of the arrival of goods to the places of demand.

In 2020, the spread of the coronavirus (COVID-19) pandemic imposed new trends in the maritime transport such as the design of a sourcing chain. In 2022, the Russian–Ukraine war imposed more challenges facing global supply chains. Thus, wide-ranging policy implications have been imposed to cope with global disruptions caused by the pandemic and with the consequences of the ware, such as restructuring the processes and organizations [11].

Central features of NPM include structural devolution, disaggregation, decentralization, competition management, and privatization. However, the key objectives of all these reforms are to improve government effectiveness and efficiency. An amplified interest in employing indicators that are used in private organizations was evident in the usage of NPM in public administration practices [12].

Accordingly, NPM mechanisms and elements are what a maritime sector needs in order to launch an administrative reform based on "strategy performance" which enhances public accountability and good governance. The NPM mechanisms and central features cannot be present during the absence of a solid organizational structure [13], [14] which should be reevaluated by any government that is due for an administrative reform as their starting point.

In designing an organizational structure of project, the organizational structure was divided by De Paula et al. [15] into levels, including (i) strategic level that is responsible for planning and the identification of the executives; (ii) direction level that is responsible for driving within the rules and limitations; and (iii) management and delivery levels that is responsible to establish the front conduction of the project. It was clarified by Zighan et al. [1] that adjusting an organizational structures requires collaboration, flexible work flows, decentralized decision-making, control, and governance. It can be claimed that the suitable organization structure supports the management of both information technology and innovation [16].

In the maritime literature, the organizational structures are not widely discussed; on the contrary, the focus was on different aspects, including the assessment of the organizational effectiveness [17], the evaluation of the corporate governance effectiveness on financial performance [18], studying the correlation between the organizational culture and strategy [19], the adoption of big data analytics in maritime organizations [20], examining the relationship between the human resources management practices and the employees performance [21], and the relationship between organizational structure and knowledge management [22]. It could be presumed that an OS depends on the division of exercises to accomplish the execution of recently arranged exercises, where the capacities, commitments, and authority of the individuals are resolved [23]. There is no particular type of OS design. It is imperative to plan OS dependent on a few elements including artifact, size, innovation, culture, climate, property, and force.

It was argued by Güler [24] that ports' structures should consider the comprehensive maritime policy, safety measures, and ecological, environmental and physical matters. Also Tagowa [25] claimed that the maritime transport organizational structure in Nigeria should have a relationship with other private and public bodies such as culture, tourism and maritime transport. It was discussed by Rocha [26] how a transport should be organized in Portugal through three levels: (i) local transport authorities that are responsible for regional transport planning, (ii) regional transport authorities that are responsible for connecting local transport authorities and mobility of a population, and (iii) national transport authority that is responsible for the development and management of national transport projects.

Both Free and Hecimovic [27] and Akpinar and Ozer-Caylan [28] argued that those reforms can be due to the need for managing their operations under harsh competitive regulations in an ever-changing business climate with unpredictable future prospects. The maritime industry exhibits unpredictable market dynamics that are influenced by even minor global developments. For example, the introduction of "Industry 4.0" presented a new "environmental change" trend for the wide range of maritime transport. On the other hand, ports can enhance their performance and competitiveness through developing laboratories and related-research activities in their structures [29].

To deal with all of these newly introduced issues in the 21st century, corporate strategies that apply a generic hypothetical context of "performance strategy" to the profitability of maritime transport are becoming increasingly relevant [30]. This further demonstrates that the marine industry has evolved into a more "business-like" environment, necessitating the development of maritime-specific strategic theories and frameworks.

The maritime transport industry is facing dynamic changes that require a new organization of work and management [31]. Thus, introducing reform in the maritime sector; need to be regarded as a part of the holistic strategic governmental administrative reform.

Thereof, the motivation of this research is to introduce an integrated maritime transport organizational structure (MTOS), and a port authority organizational structure (PAOS). A well-developed and sustainable OS is considered one of the most important factors of administrative reform and development where it can clarify tasks, relationships, and methods of communication. The proposed MTSO and PAOS help to provide the necessary planning and organizing administration of the maritime transport industry and to ensure the ability of public accountability and governmental governance introducing a better environment for service introduced and higher customer satisfaction.

The upcoming sections of this paper are organised as: Section 2 is about the scope of this research, while the research questions are shown in Section 3. The research methodology used in this paper and research validity will be explained in Section 4. In Section 5 the characteristics and challenges facing current MTOSs and PAOS will be discussed. Section 6 presents the proposed maritime transport organizational structure in addition to the criticisms of experts and academics about the newly proposed model. The final model and its criticisms will be presented in Section 7. Finally, Section 8 presents the conclusion, research validity, and the future research take place.

2 RESEARCH SCOPE

This research conducts a case study in one Arab country aiming to assess the proposed OS in a country that shares common history, similar religious background, similar beliefs and similar political conditions with other countries within the same geographical area. This can be one step towards future generalizability.

3 RESEARCH QUESTIONS

According to the discussion in Section 1, research questions have been addressed as follows: (1) What are the most appropriate elements needed in a maritime transport organizational structure in Arab countries? (2) What are the most appropriate elements needed in a port organizational structure in Arab countries? And, (3) How can these elements integrate to show an organizational structure that can help in creating a high performing organizational environment?

4 RESEARCH METHODOLOGY

This research adopts the qualitative, descriptive, and exploratory constructive research that explores the phenomenon of organizational structures in ports and maritime systems in order to describe the problem in the MTOSs. The methodology is shown through the following steps:

i. The research is a qualitative one that depends on exploring literature of governance, governmental performance and accountability; in addition to exploring several elements in 47 world MTOSs that are selected according to six criteria shown in Section 6.1.
ii. The data collection method used will be through interviewing (semi-structured/ unstructured) academic international maritime experts with not less than 15 years of experience in addition to governmental officials of not less than 5 years of experience in the governmental practical field. Furthermore, data and information of 47 MTOSs will be collected using documentation from official websites of 47 governments;
iii. In accordance to the data collected from both literature and interviews, the authors will be able to describe the number of departments needed, the activities, roles and responsibilities in addition to characteristics and challenges facing current MTOSs. Consequently they can propose the intended MTOS.
iv. The proposed structure will be introduced through the purposive expert sampling technique to academics and experts (as shown in Section 6.3) for validation.
v. After validations and amendments, the final OSs will be implemented as a pilot study in one Arab country in order to examine the constructive applied research of the paper (Section 7).
vi. Evaluation of quality of research according to Yin [32] will be discussed in Section 8.

5 CHARACTERISTICS AND CHALLENGES FACING THE MTOS

This section introduces the characteristics and challenges facing the MTOSs in Section 5.1 and the characteristics of the international MTOSs in Section 5.2 in order to be able to construct a new OS.

5.1 Characteristics and challenges facing current maritime OS

The discussion in Section 1 shows that various maritime transport organizational structures have a set of challenges. Some MTOSs have lack of clarity of the organizational structures where there is conflict in some specialties between departments and entities. In addition, there is a separation between the maritime administration and the maritime authority, non-integration with specialized seaports, and absence of some important departments and sections that enhances the maritime transport industry.

Other MTOSs are suffering from lack of clarity in the departments concerned with training workers, developing their capabilities, and raising their efficiency. Also, other challenges appear in the absence of value-added logistics activities, lack of integration with the economic and free trade zones, and challenge to keep up to date with international agreements and conventions, either in terms of implementation or continuous follow up on their amendments for full and continuous implementation.

The above challenges affect the MTOSs where there is limited ability to attract the required volume of seaborne trade, or the expansion of re-export trade and the benefit from their distinguished geographical locations, as well as the navigation services that can be provided. Moreover, the challenge of following up the technological development in the maritime field, like digitization, IoT, Big Data, etc., that affects the efficiency of the performance of maritime transport components.

5.2 International MTOSs characteristics

Evaluating a number of the international maritime transport organizational structures show that there is a great variation between OSs. Some international MTOSs have entities that have similar missions and activities with others reinforces, which introduce the idea of the "multiple hands problem". Also, some central departments in the structure follow public administrations, the chain of operations flow is not clear in the current organizational structures, and the Authority for Maritime Safety is not represented in some OSs.

In addition, the presence of a large number of central administrations in the structures is not consistent with the principles of modern administration in the decentralization approach. The Department of International Treaties and Agreements is not clear in some OSs. And, the data bank, the central statistics department, and the accounting agency report analysis unit are not represented in some OSs.

6 PROPOSED MARITIME TRANSPORT AND SEA PORT AUTHORITY ORGANIZATIONAL STRUCTURES

This section reviews the international maritime organizational structures. This is to analyse the main common functional areas that will be used in developing the proposed OSs as shown in Sections 6.1 and 6.2

6.1 International maritime organizational structures

MTOSs for 47 countries were selected by the authors to review their organizational structures they were selected from all continents and regions of the world such as China, Singapore,

Japan, UAE, Qatar, India, USA, Canada, UK, Russia, Spain, Belgium, Greece, Romania, Turkey, Mexico, Peru, Ecuador, Morocco, South Africa, etc. The selection of the MTOSs was based on nine criteria which are: top LSCI, last 10 LSCI, handling rates, time at ports, regional connectivity, ship building, ship ownership, ship registration, and ship scrap. It becomes obvious that some OSs have many maritime-related activities and functions while another OSs have a limited focus on certain activities. This is due to the strategic maritime policy of the country, trade mechanism, and the availability of the industry requirements such as infrastructure. China and South Korea, for example, have OSs that comprise most of the maritime-related activities and functions while other countries such as Panama are specialized in a specific activity like ship registration.

The number and type of functional areas differ from one OS to another. The analysis of the selected OSs shows that some functional areas are frequent in most of the OS. Safety, security and marine services have the highest priority in the selected OSs as shown in Fig. 1, followed by legal issues and international relations.

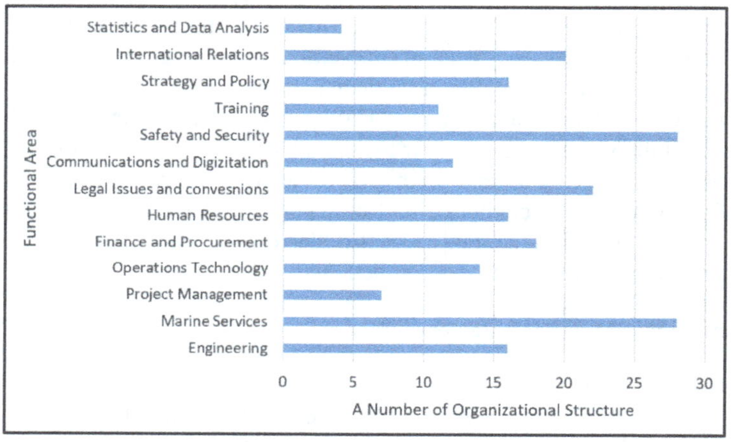

Figure 1: The functional areas of the selected organizational structures.

Strategy and policy, engineering services, and finance and procurement receive adequate consideration in the OSs. However, statistics and data analysis and project management received the less reflection in the OSs. On the other hand, the most OSs are located in Asia, where the majority of the maritime-related functional areas, like Japan, China, Singapore and South Korea, exist.

6.2 The proposed MTOS

The proposed MTOS (Fig. 2) is considered to be a restructuring of the current most common MTOSs as revealed from the analysis of data from Section 6.1, where the work undertook re-arrangement of the departments and sections in the various public bodies and departments in addition to trying to adhere to the requirements of the International Maritime Organization (IMO) and adding some new departments and sections to support the activity of national MTOS.

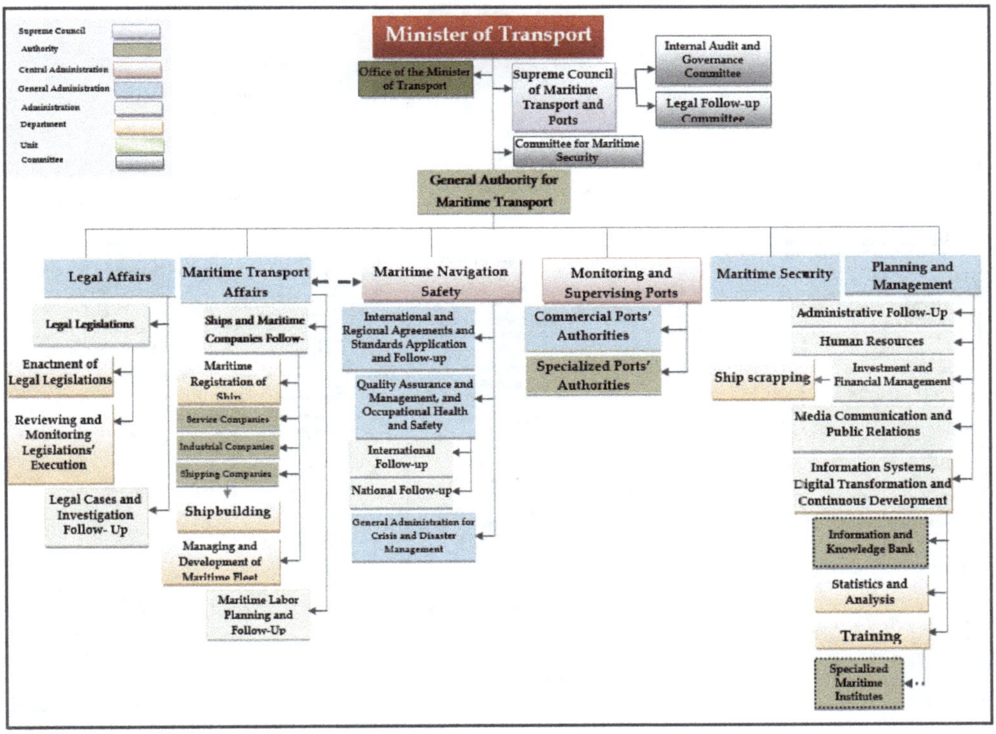

Figure 2: The proposed MTOS.

6.3 Purposive expert sampling technique validation of the proposed MTOS

This section is about validating the proposed OS. The proposed MTOS was validated by maritime officials and senior academic staff members in the maritime field as mentioned in Section 4. They ensured that a complete reengineering is crucial for better accountability and governance of maritime transport in Arab countries and that the proposed OS does not introduce the needed reengineering.

They stated the importance of having a general authority for each central task according to the IMO which are (1) maritime safety; (2) maritime excellence; and (3) ports. Accordingly, the authors made the needed amendments and introduced the final MTOS to the same validation committee that reviewed the first proposal. They approved of it (Sections 7.1 and 7.2) and recommended implementation in at least one port in the Arab countries for more validation.

7 FINAL PROPOSED MARITIME TRANSPORT AND SEA PORT AUTHORITY ORGANIZATIONAL STRUCTURES

The proposed MTOS (Fig. 3) is considered to be a reengineering of the selected MTOSs, where it radically redesigns the processes, departments, and divisions to achieve the necessary development in performance measures.

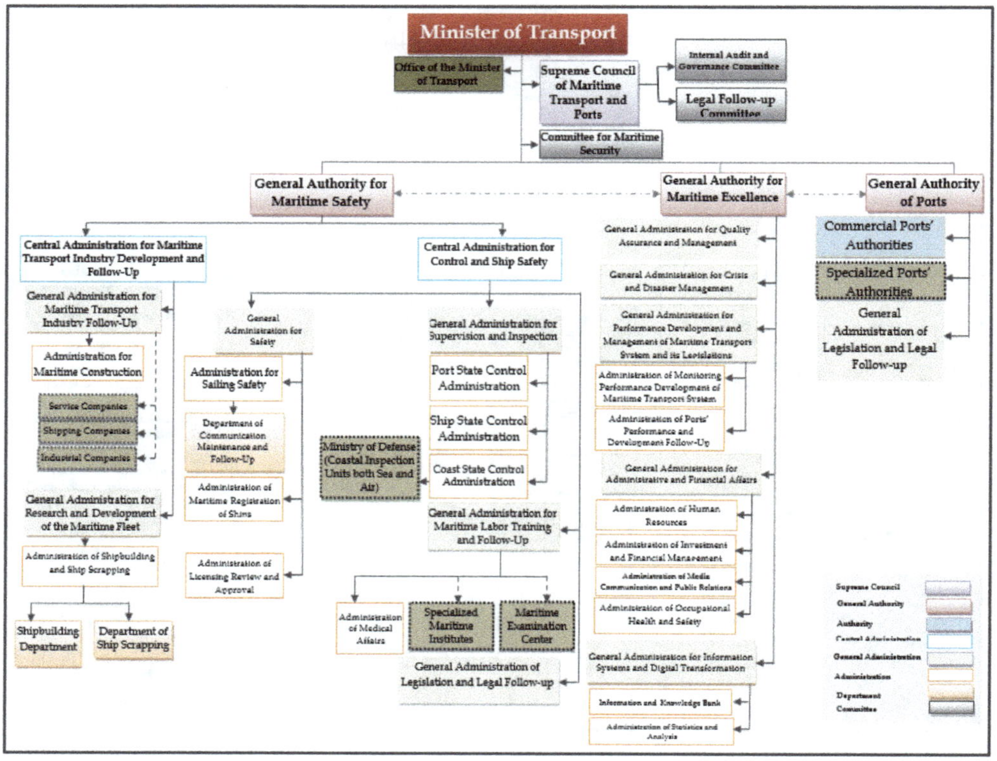

Figure 3: The final proposed MTOS.

7.1 The final proposed MTOS

From this standpoint, the final proposed OS contained the following key elements introduced in Fig. 3:

1. The General Authority of Ports: proposed to ensure the decentralization between several port authorities within a state and to give freedom to each port authority manager to deal with the inputs and conditions of his/her authority.
2. General Authority for Maritime Excellence: proposed due to the growing importance of the institutional excellence in the world, especially with the increasing risks and changes in the environment surrounding business.
3. General Authority for Maritime Safety: works in cooperation and coordination with the General Authority for Ports which has the safety of ports as one of its responsibilities. As well as the management and development of any manufacturing or business-related activities conducted by the maritime transport system.

7.2 The proposed general authority of ports OS

The term “port authority” is widely used to refer to any of the terms responsible for the functions of construction, management, and operation of port facilities. The proposed General Authority of Ports OS shown in Fig. 4 is the validated final version after applying all amendments, comments and modifications recommended by the interviewees as discussed in Section 6.3.

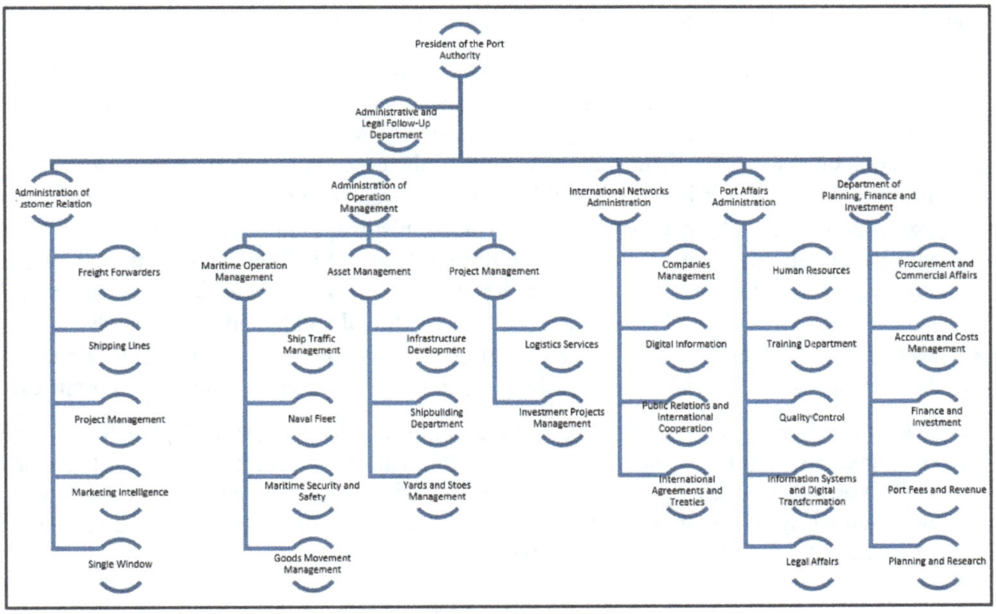

Figure 4: The proposed seaport authority OS.

The General Authority of Ports has different forms that differ from one country to another, and from one port to another within the same country.

7.3 Implementation of proposed model

The proposed OSs have been submitted and presented to the Egyptian Public Authority for Transport Planning Administration in the Ministry of Transport, for review and validation. After several meetings with maritime officials and experts in the Ministry, the proposed OSs have been approved and escalated to the Minister for consideration in the future.

Furthermore, the researchers received an official letter from the Egyptian Public Authority for Transport Planning Administration in the Egyptian Ministry of Transport regarding the start of implementation of the final proposed OS and the predicted effectiveness when fully implemented.

8 CONCLUSION, RESEARCH VALIDITY, AND THE FUTURE RESEARCH

This research aims at studying and analysing the organizational structures of several maritime transport bodies in different countries, identifying the current challenges, and in turn, proposing a validated MTOS and one PAOS.

Hence, this research highlighted the trend towards decentralization in decision-making encouraging decision-making at various administrative levels. This can be achieved by establishing detailed and clear organizational structures for all the main bodies operating in maritime transport to ensure the smooth flow of orders and the absence of conflicts of jurisdiction or duplication of work. The proposed OS was validated and approved through interviews with both experts and academic staff in the field of maritime transport the validation of the proposed OSs as discussed in Sections 6 and 7.

A number of criteria were set by Yin [32] according to which the quality of a case study can be evaluated. These criteria are discussed in relation to this research as follows:

- Construct validity: A number of researches and theories were extracted from literature and various references presented in Sections 1 and 5. They were used to identify the importance of research and the problem faced by Maritime Transport authorities in several countries. This ensures the existence of the construct validity in the research.
- Internal validity: This is not applicable as the research is an exploratory one.
- External validity: Arab countries share similar history, similar religious background, similar beliefs and similar political conditions [33]–[35]. The implementation of the proposed model in one Arab country with promised effectiveness as discussed in Section 7.3. This gives a probability of existence of the external validity in the research.
- Reliability: The research is a single case study; and to approve reliability it will need to be repeated using the same steps as those used in this research in other Arab countries.

For future works, it is recommended to develop a clear and detailed plan for change management before implementing any administrative reform that would disturb employees such as changing the organizational structure or amending job description while adhering to full transparency about any suggested alternation to avoid resistance to change and the accompanying problems.

REFERENCES

[1] Zighan, S., Alkalha, Z., Bamford, D., Reid, I. & Al-Zu'bi, Z.M.F., Servitisation through structural adaptation. *Journal of Service Theory and Practice*, **31**(3), pp. 468–490, 2021. DOI: 10.1108/JSTP-06-2020-0144.

[2] Swee, C.G., Making performance measurement systems more effective in public sector organizations. *Measuring Business Excellence*, **16**(1), pp. 31–42, 2012.

[3] Newcomer, K. & Caudle, S., Public performance management systems: Embedding practices for improved success. *Public Performance and Management Review*, **35**(1), pp. 108–132, 2011.

[4] Mulgan, R., Accountability: An ever-expanding concept? *Public Administration*, **78**(3), pp. 555–573, 2000.

[5] Osborne, S.P., *Public Management: Critical Perspectives*, Routledge, 2002.

[6] Denhardt, J.V. & Denhardt, R.B., *The New Public Service: Serving, Not Steering*, Routledge: New York, 2011.

[7] Hansson, L. & Longva, F., Contracting accountability in network governance structures. *Qualitative Research in Accounting and Management*, **11**(2), pp. 92–110, 2014.

[8] Islam, A.E., Investigating governmental (public sector) performance management in Egypt. Unpublished PhD thesis, University of Strathclyde, Glasgow, 2010.

[9] Islam, A.E., Towards governmental administrative reform: Organizational charting in Egypt. *Journal of Faculty of Commerce for Scientific Research*, **52**(2), Part I, pp. 1–33, 2015.

[10] UNCTAD, Review of maritime transport 2020. https://unctad.org/system/files/official-document/rmt2020_en.pdf. Accessed on: 19 Apr. 2021.

[11] Tortorella, G., Narayanamurthy, G., Godinho Filho, M., Portioli Staudacher, A. & MacCawley, A.F., Pandemic's effect on the relationship between lean implementation and service performance. *Journal of Service Theory and Practice*, **31**(2), pp. 203–224, 2021. DOI: 10.1108/JSTP-07-2020-0182.
[12] Buffat, A. & Larpin, B., The impact of administrative reforms on democratic governance in Switzerland. *Administrative Reforms and Democratic Governance*, eds J. Eymeri-Douzans & J. Pierre, Routledge: London, 2011.
[13] Islam, A.E., Accountability role in the Egyptian revolution. *Perspectives on Performance*, **10**(3), pp. 18–25, 2013.
[14] Islam, A.E & Khourshed, N.F., Placing performance measurement and performance management in context of strategic management process: A conceptual framework. *Proceedings of the 8th Conference on Performance Measurement and Management Control*, 30 Sep.–2 Oct., Nice, France, 2015.
[15] De Paula, A.B.D.O.N., Leiras, A. & Lukosevicius, A., Organizational structure for disaster management projects. *WIT Transactions on The Built Environment*, vol. 173, WIT Press: Southampton and Boston, pp. 33–42, 2017.
[16] Chapman, R.L., Soosay, C. & Kandampully, J., Innovation in logistic services and the new business model: A conceptual framework. *Managing Service Quality: An International Journal*, **12**(6), pp. 358–371, 2002. DOI: 10.1108/09604520210451849.
[17] Cetin, C.K. & Cerit, A.G., Organizational effectiveness at seaports: A systems approach. *Maritime Policy and Management*, **37**(3), pp. 195–219, 2010.
[18] Giannakopoulou, E.N., Thalassinos, E.I. & Stamatopoulos, T.V., Corporate governance in shipping: An overview. *Maritime Policy and Management*, **43**(1), pp. 19–38, 2016.
[19] Shin, S.H. & Shin, Y.J., The impact of organizational culture and strategy on shipping liner's awareness and utilization of the 4th IR technologies. *Maritime Policy and Management*, **47**, pp. 1–19, 2020.
[20] Zhang, X. & Lam, J.S.L., A fuzzy Delphi-AHP-TOPSIS framework to identify barriers in big data analytics adoption: Case of maritime organizations. *Maritime Policy and Management,* **46**(7), pp. 781–801, 2019.
[21] Aydoğan, E. & Arslan, Ö., HRM practices and organizational commitment link: Maritime scope. *International Journal of Organizational Analysis*, **29**(1), pp. 260–276, 2021.
[22] Mahmoudsalehi, M., Moradkhannejad, R. & Safari, K., How knowledge management is affected by organizational structure. *The Learning Organization*, **19**(6), pp. 518–528, 2012.
[23] Galván, I.P., *Designing Local E-Government: The Pillars of Organizational Structure*, Emerald Publishing: Bingley, 2019.
[24] Güler, N., Black Sea maritime transport corridors in relation with EU projects. *WIT Transactions on The Built Environment*, WIT Press: Southampton and Boston, p. 79, 2005.
[25] Tagowa, W.N., The Changing patterns of the boat regatta as maritime heritage in Nigeria. *WIT Transactions on The Built Environment*, WIT Press: Southampton and Boston, p. 79, 2005.
[26] Rocha, N.G., A transport strategy for Portugal. *WIT Transactions on The Built Environment*, vol. 111, WIT Press: Southampton and Boston, pp. 3–14, 2010.
[27] Free, C. & Hecimovic, A., Global supply chains after COVID-19: The end of the road for neoliberal globalization? *Accounting, Auditing and Accountability Journal*, **34**(1), pp. 58–84, 2021.

[28] Akpinar, H. & Ozer-Caylan, D., Managing complexity in maritime business: Understanding the smart changes of globalization. *Competitiveness Review*, in print. DOI: 10.1108/CR-10-2020-0128. Accessed on: 25 Apr. 2021.
[29] Russo, F. & Rindone, C., Structural factors for a third-generation port: Planning interventions for research and development in Gioia Tauro, Italy. *Ten-T Node. Urban and Maritime Transport XXVII*, p. 204, 2021.
[30] Wang, P. & Mileski, J., Strategic maritime management as a new emerging field in maritime studies. *Maritime Business Review*, **3**(3), pp. 290–313, 2018.
[31] Jugović, A., Mezak, V. & Lončar, S., Organization of maritime passenger ports. *Pomorski Zbornik*, **44**(1), pp. 93–104, 2006.
[32] Yin, R.K., *Case Study Research*, 3rd ed., SAGE: Thousand Oaks, 2003.
[33] Barakat, H., Arab identity. *The Arab World: Society, Culture and State*, University of California Press: Berkeley, LA, 1993. https://teachmideast.org/articles/facets-of-arab-identity-halim-barakat/. Accessed on: 30 Mar. 2022.
[34] Islam, A.E. & Labib, A.A., Performance measurement and culture in Arab countries. *1st Annual Conference, Total Quality Management in Service Industries (TQMISI)*, Alexandria, Egypt, 20–21 Jan. 2007, 2007.
[35] Ali, A.J., Organizational development in the Arab world. *Journal of Management Development,* **15**(5), pp. 4–21, 1996.

MAIN CAUSES OF CARGO FIRE INCIDENTS IN RO-RO SPACES: AN ANALYSIS OVER THE LAST 25 YEARS

ÁFRICA MARRERO*, PACO GASPARÍN†, FRANCISCO RODERO‡ & PABLO SANZ
CENIT (RTD Group of CIMNE-CERCA), Spain

ABSTRACT

The FIRESAFE study from EMSA revealed that 90% of all ro-pax ship fires are initiated in the carried cargo (vehicles and cargo units), which can be everything from brand new to poorly maintained, rebuilt, or unsafe. Today, all cargo (except dangerous goods) is loaded without consideration to the hazards they pose, leaving much room for fire prevention. This paper focuses on the analysis of available historical data related to fire accidents produced inside ships. The objective is to find out what the main causes and their origins are. The methodology includes the following steps: (1) the compilation of information from several maritime sources, (2) the corresponding homogenization based on preliminary outcomes of LASH FIRE (H2020 funded project, Grant Agreement #814975, in which this publication is framed), (3) a brief overview on the design of the underlying database which supports the analysis and finally, (4) the main conclusions. The analysis considers not only the type of ship and cargo (classified as dangerous or non-dangerous goods, if it is transported in a vehicle or not and, in that case, what type of vehicle, etc.) but also the ro-ro space where it was located. The resulting information is used to identify patterns in the root causes (electrical, mechanical, overheating, etc.) in order to create knowledge that can be further used for the development of innovative solutions for the implementation of fire hazard management.

Keywords: ignition prevention, risk analysis, fire safety, stowage.

1 INTRODUCTION

Safety is a key parameter in any topic, and maritime transport is not an exception. Nowadays, there is an increasing concern from the International Maritime Organization (IMO) and other regulatory bodies to improve safety, especially on reducing the risk of fires on board ro-ro ships. Proof of this is that in 2019 IMO's sub-committee [1] held a meeting with the aim of creating guidelines focused on fire safety on ro-ro vessels

Transportation of ro-ro cargo entails lot of fire hazards for the vessel because vehicles have several elements such as the engine, fuel tanks or electrical systems that could cause an ignition [2]. FIRESAFE study showed that 30% of fires in ro-ro ships are originated on vehicle decks and 90% of these were originated in the cargo transported [3].Therefore, ro-ro ships need special attention regarding to fire safety. Besides this, these types of vessels lack compartmentalization of cargo space, which means that in case of fire on a cargo unit, the entire deck is potentially threatened.

However, current cargo placement in ro-ro spaces is not optimal against fire hazards. There have been several fire incidents in recent years, originated mainly in those vehicles transporting the cargo [4].These kinds of incidents concern not only from the economic loss perspective but also the risk they represent for human beings.

In front of this scenario, this paper describes and evaluates the main causes of cargo unit fires over the last 25 years. The work is presented as the first step to solve the problem. To do so, it is necessary to know what is being transported and the risks involved in cargo units.

* *ORCID: https://orcid.org/0000-0003-2949-5169*
† *ORCID: https://orcid.org/0000-0003-3064-7764*
‡ *ORCID: https://orcid.org/0000-0002-9613-2993*

WIT Transactions on The Built Environment, Vol 212, © 2022 WIT Press
www.witpress.com, ISSN 1743-3509 (on-line)
doi:10.2495/UMT220141

Cargo other than dangerous goods have not been regulated so far; no recommendations on how they should be transported either. This situation drives to operator-specific standards for a safer transport based on their own experiences.

This article aims at analysing historical data and identifying patterns, main causes and problems derived from the lack of information about the loads to be transported to establish a knowledge baseline of the problem for future solutions preventing cargo fires on ro-ro spaces. The approach of the methodology used involves the following steps:

1. Since needed information is not available as required, firstly is the search for different sources, to analyse them to gather as much relevant information as possible and discarding and filtering all unnecessary information. At this point, all the data is formatted in a spreadsheet (user-friendly) and the process is done manually.
2. Transformation to a structured format oriented to speed up the access of the information by other software components (machine-friendly).
3. Once the cargo fire risk database has been finalized, it is analysed considering accident factors such as main origin, causes, where it occurred (type of deck: open, closed or weather) or year among other relevant characteristics.

2 DATA COLLECTION

In order to obtain an accurate and realistic analysis based on historical data, it is of paramount importance to collect all incidents that could be relevant and to create a solid background. Thus, an intense work has been done looking for and gathering incidents, accidents or reports coming from several heterogeneous sources of information.

Data selected must fulfil two conditions: (a) the fire origin must be located in a ro-ro space, and (b) the fire must be started in the vehicle or the cargo transported by (i.e. fires started in the engine room of the ship are discarded).

Sources include public and private organizations, ship operators, research institutes and journals; all of them are briefly described below:

1. FSI 21/5: Document provided by IMO [5] which offers a wide list of maritime accidents in ro-ro passenger ships from 1994 to 2011.
2. GISIS: Global Integrated Shipping Information System [6] is an information system provided by IMO that, among other services, provides a collection from several marine casualties and incidents.
3. Ship operators: The representatives from the Maritime Operators Advisory Group (MOAG) of the LASH FIRE project [7], gave access to incidents or near-miss related information in the ro-ro spaces from their own databases.
4. ForeSea [8]: Information system used in Sweden and Finland to record accidents, incidents or near misses at sea.
5. Official organizations in Mediterranean countries: Research in official organizations from Mediterranean countries such as CIAIM (Comisión Permanente de Investigación de Accidentes e Incidentes Marítimos) [9], HBMCI (Hellenic Bureau for Marine Casualties Investigation) [10] or MTIP (Ministry for Transport, Infrastructure and Capital Projects) [11] has been performed to complement the rest of sources, where only northern locations were found.
6. EMCIP: The European Marine Casualty Information Platform [12] is a database and a data distribution system, also connected to GISIS, operated by EMSA (European Maritime Safety Agency), the European Commission and the EU/EEA Member States.
7. Other sources: Apart from the above, several sources such as journals (e.g. *Journal of Maritime Research*) or private companies' reports have been analysed. The main

purpose of these sources was to complete unknown information from incidents already recorded although they also provided some new incidents.

Project-related sources were also included: outcomes from an internal workshop (several marine casualties in ro-ro spaces, some of them already covered by FSI and GISIS but contributing with additional details) or internal fire investigation database from RISE, the Research Institutes of Sweden [13], who plays the coordinator role in the Consortium of LASH FIRE project.

At the end of the process, about 230 records have been obtained; existing duplicates have been considered only once and have been assigned to that source where more detailed and accurate information was found (Fig. 1).

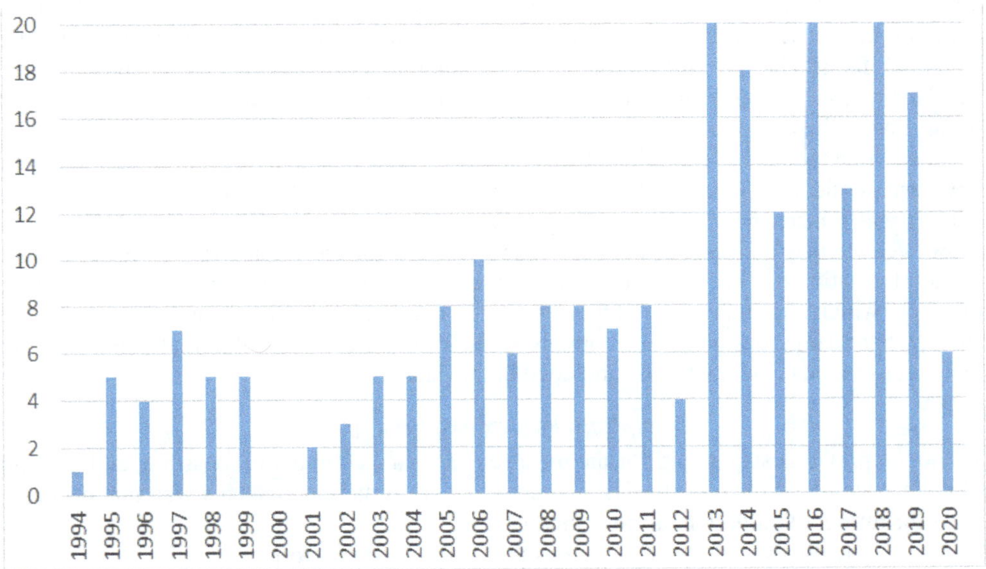

(a)

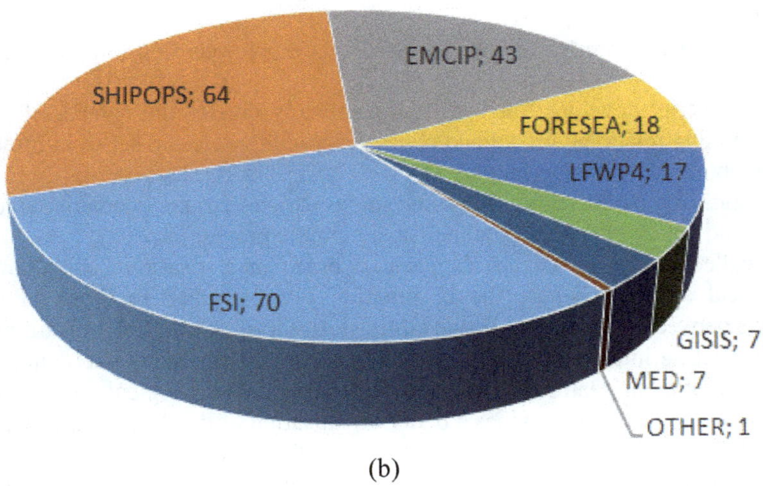

(b)

Figure 1: Summary of collection. (a) Yearly distribution; and (b) Incidents per source.

Nevertheless, this work is not limited to gathering records but also in classifying them with respect to the fire origin and cause, and merging the information provided from different sources to perform a single and consistent database.

The first classification variable is the fire origin, key element to perform a risk-based analysis, as probability of ignition occurrence may vary significantly. The analysis distinguishes between refrigeration units (commonly referred as reefers), conventional vehicles (trucks, cars and buses), special vehicles (includes tractors, forest vehicles, …), new energy carriers (those vehicles which use alternative fuels, including electricity) and, due to its specific properties, dangerous goods, regardless of not being a vehicle.

The second level of classification is the cause of the fire, which gives information about the factors that lead to an increase of the probability of ignition. Detecting these conditions allows considering them as potential-ignition factors in the risk analysis. Some of the most common fire causes are electrical faults in cables or connections (especially for reefer units), electrical faults (in the unit, cabin or in the engine compartment), mechanical faults, overheating (in the chassis or in the engine compartment), engine leakage of flammable liquid or ignition in the cargo itself.

In addition to the abovementioned, up to almost 20 relevant data fields from each incident have been collected like the date and location of the incident, the severity, what goods the vehicle was transporting (if any), what goods were transported by nearby vehicles or if the origin vehicle was connected (and charging) when the ignition occurred. For those reports, extra information has been gathered such as the type of vessel (ro-ro, ro-pax or car-carrier), the deck (which one and type) or if the vessel was loading, on passage, or unloading. In those cases, where dangerous goods were involved, their IMDG code (The International Maritime Dangerous Goods Code) was also included in the database.

3 DATABASE CONSTRUCTION

While the previous step aimed for the most relevant data compilation possible, at this stage of the methodology the objective is to structure the information in such a way that becomes agile for further exploitation, as described later.

More in depth, the next list enumerates the main actions performed:

1. Using a macro developed with Visual Basic for Applications, a new tab is filled with all the data in a tabular way. The objective is to prepare the data to be used to populate the final database.
2. A second macro exports the tab generated in the previous step to a comma separated values file.
3. Finally, a script developed in Python dynamically creates the database based on the corresponding relational model and populates it with a selected set of information from the file obtained in the previous step.

Fig. 2 graphically depicts this process and shows which steps are interactive (requires user interaction) and which are not (just launch an automatic process).

The so-called cargo fire hazard database is based on a relational model, which is a methodological approach where the information is represented in terms of tables and relationships between these tables. While tables store entities or logical units of information, relationships define interaction or association between tables; each table contains a set of attributes, each of them, featuring a relevant characteristic of the logical unit of information.

Such a structure not only provides a better organisation of the data but a way to speed up queries against the database in order to retrieve information that can be used for many purposes. Queries can be performed against one or more tables, extracting a final set of records composed of logical and mathematical operations between involved attributes.

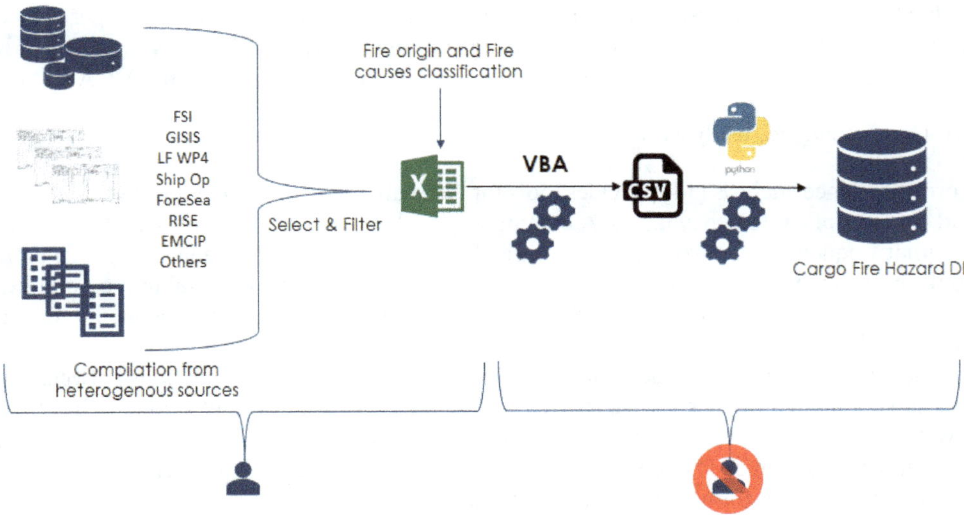

Figure 2: Workflow of the methodology (data collection and database construction).

3.1 Implementation

Using a relational model does not imply a unique implementation because there are lot of both free and commercial solutions, each of them offering different options in terms of prices, licensing, portability, scalability or technical support. In the context of the project where these tasks have been developed, the following requirements, among others, were taken into consideration:

- Development frameworks, programming languages and protocols, as mainly examples, must be free and open, if possible.
- Availability in the most common operating systems and easy to install and/or deploy, ensuring portability.
- Products fast to learn for future maintenance purposes by third-party staff.

Finally, the selected platform is composed of Python as programming language and Sqlite as the database management system (DBMS). Python is an interpreted, high-level, general-purpose object-oriented programming language emphasizing code readability for large-scale projects, including a comprehensive standard library. Sqlite is a relational database management system (RDBMS) which, in contrast to many other systems, it is not a client-server-based engine. It implements most of the SQL (structured query language) standard while ACID (atomicity, consistency, isolation and durability) properties of database transactions are guaranteed. Its source code is in the public domain and free of charge to everyone to use for any purpose.

4 METHODOLOGY OF DATA ANALYSIS

From the database constructed, an exploratory analysis is carried out to select the variables to be used to apply the multiple correspondence analysis. Once applied, the different correlations or patterns are searched for/analysed to find out the relationships between the variables.

R Studio software (version 4.1.0) is used for the analysis. The MCA has been performed using the MCA (multiple correspondence analysis) function (FactoMineR package). The R package factoextra is then used to produce a ggplot2 based visualisation of the MCA results.

4.1 Correspondence analysis (CA)

Correspondence analysis [14] provides tools for the analysis of the associations between rows and columns of contingency tables. A contingency table is a two-entry frequency table where the joint frequencies of two qualitative variables are reported. For instance, a (2 × 2) table could be formed by observing, from a sample of n individuals, two qualitative variables (in general they are *s* variable). The table reports the observed joint frequencies. In general ($n \times s$) tables may be considered.

The main idea of CA is to develop simple indices that will show the relations between the row and column categories. These indices will define what is the weight a column has for a row category and vice versa. CA is also related to both issues of reducing the dimension and decomposing the table. The idea is to extract the indices in decreasing order of importance so that the main information of the table can be summarised in spaces with smaller dimensions.

Contingency tables are very useful to describe the association between two variables in very general situations. The two variables can be qualitative, in which case they are also referred to as categorical variables. Each row and each column in the table represents one category of the corresponding variable. The entry n_{ij} in the table N (with dimension ($n \times s$)) is the number of observations in a sample which simultaneously fall in the *i*th row category and the *j*th column category, for $i = 1, \ldots, n$ and $j = 1, \ldots,$ s. For notational simplicity [15] the matrix is first converted to the correspondence matrix $\boldsymbol{P}$ by dividing $\boldsymbol{N}$ by its grand total $n = \sum_i \sum_j n_{ij} = 1^T N 1$ (the notation 1 is used for a vector of ones of length that is appropriate to its use; hence the first one is $n \times 1$ and the second is $s \times 1$ to match the row and column lengths of $\boldsymbol{N}$). Correspondence matrix:

$$P = \frac{1}{n} N. \tag{1}$$

The following notation is used:

- Row and column masses:

$$r_i = \sum_{j=1}^{n} p_{ij}, \qquad c_j = \sum_{i=1}^{s} p_{ij}, \tag{2}$$

$$i.e., \ r = P1, \qquad c = P^T 1.$$

- Diagonal matrices of row and column masses:

$$D_r = diag(r) \quad \text{and } D_c = diag(c). \tag{3}$$

Note that all subsequent definitions and results are given in terms of these relative quantities $P = \{p_{ij}\}$, $r = \{r_i\}$ and $c = \{c_j\}$, whose elements add up to 1 in each case. Multiply these by n to recover the elements of the original matrix N: $np_{ij} = n_{ij}$, nr_i = *i*th row sum of N, ncj = *j*th column sum of N.

The computational algorithm to obtain coordinates of the row and column profiles with respect to principal axes, using the singular value decomposition (SVD), is as follows (eqns (4)–(10)):

Step	Description	Equation	
1	Calculate the matrix S of standardized residuals	$S = D_r^{-\frac{1}{2}}(P - rc^T)D_c^{-1/2}$	(4)
2	Calculate the SVD of S	$S = UD_\alpha V^T$ where $U^TU = V^TV = I$ where $D\alpha$ is the diagonal matrix of (positive) singular values in descending order: $\alpha 1 \geq \alpha 2 \geq$.	(5)
3	Standard coordinates Φ of rows	$\Phi = \mathrm{D}_\mathrm{r}^{-\frac{1}{2}}V$	(6)
4	Standard coordinates Γ of columns	$\Gamma = D_c^{-\frac{1}{2}}V$	(7)
5	Principal coordinates F of rows	$\mathrm{F} = \mathrm{D}_\mathrm{r}^{-\frac{1}{2}}UD_\alpha = \Phi D_\alpha$	(8)
6	Principal coordinates G of columns	$\mathrm{G} = \mathrm{D}_\mathrm{c}^{-\frac{1}{2}}VD_\alpha = \Gamma \mathrm{D}_\alpha$	(9)
7	Principal inertias λk	$\lambda_\mathrm{k} = \alpha_k^2, \quad k = 1, 2, \ldots, K$ where $K = \min\{n-1, s-1\}$	(10)

The rows of the coordinate matrices (from step 6 to step 9) refer to the rows or columns, as the case may be, of the original table, while the columns of these matrices refer to the principal axes, or dimensions, of which there are min {n−1, s−1}, i.e., one less than the number of rows or columns, whichever is smaller.

That is, the weighted sum-of-squares of the principal coordinates on the *k*th dimension (i.e., their inertia in the direction of this dimension) is equal to the principal inertia (or eigenvalue) $\lambda_k = \alpha_k^2$, the square of the *k*th singular value, whereas the standard coordinates have weighted sum-of-squares equal to 1. All coordinate matrices have orthogonal columns, where the masses are always used in the calculation of the (weighted) scalar products.

The inertia is also the sum of squares of the singular values or the sum of the eigenvalues:

$$\text{inertia} = \sum_{\mathrm{k}=1}^{K} \alpha_k^2 = \sum_{k=1}^{K} \lambda_k. \tag{11}$$

Finally, the coordinate matrix is projected onto the principal axes where the greatest inertia accumulates. Theoretically, the first two dimensions of this matrix (1–2) are projected, but more can be projected, usually up to five axes. Then, the different combinations between them are made for their projections. Once they are done, the pertinent conclusions are drawn.

The extension of simple correspondence analysis to the case of several nominal variables (multidimensional contingency tables) is called multiple correspondence analysis and uses the same general principles as the previous technique.

4.2 Multiple correspondence analysis (MCA)

The MCA [16] is applied to contingency tables where there are (n) individuals in rows and (s) categorical variables with (pi = 1, 2, …, s) mutually exclusive and exhaustive in columns (s).

The data table has the form:

$$Z = [Z_1, Z_2, \dots, Z_s]\text{: where } Z_i \text{ is a matrix } (n \cdot p_i)\text{, such that:}$$
$$\begin{cases} 1 \text{ if the } i\text{th individual chooses modality } j \\ 0 \text{ if the } i\text{th individual does not choose modality } j \end{cases}$$

The multiple correspondence analysis is based on performing a correspondence analysis on the so-called Burt matrix: B = Z'Z

The Burt of MCA is the application of the basic CA algorithm to the matrix B, resulting in coordinates for the J categories (B is a symmetric matrix).

5 RESULTS

To begin with, the database will be quickly analysed to find out what information it contains. To do this, a summary is made, which can be seen in Fig. 3.

```
            Fire.origin.1           Fire.origin.2                 Fire.cause       Year                Severity        Type.of.vessel
Conventional vehicle:67    Truck            :39    Cargo                 : 2   2013    :11   Less serious    :11    Car carrier: 6
Dangerous goods     :10    Value            :30    Collision             : 1   2016    :10   Marine incident:30    Ro-pax     :76
New energy carrier  : 1    Car              :25    Electrical            :77   2014    : 9   Near miss       :20    Ro-ro      :14
Reefer unit         :30    RVs              : 8    Leakage               : 8   2008    : 7   Serious         : 6    NA's       :21
Special vehicle     : 9    Flammable liquid: 4    Mechanical failure    : 2   2015    : 7   Very serious    : 2
                           Bus              : 3    Overheating           :26   (Other):72   NA's            :48
                           (Other)          : 8    Spontaneous ignition: 1   NA's    : 1
           Location      Occurred.when Closed.or.open Dangerous.goods Code.dangerous.good Was.charging.
Mediterranean Sea: 9   Arrival    : 7   Closed :84     No :107         3       :  4        NA  : 4
Baltic Sea       : 7   Departure :10   Open   : 1     Yes: 10         1       :  1        No  : 2
North Sea        : 4   In port    :11   Weather: 3                     2#2,5#1:  1        Yes :25
English Channel  : 2   Loading    :11   NA's   :29                     4#1     :  1        NA's:86
Adriatic Sea     : 1   On passage:69                                  4#3     :  1
(Other)          :21   Unloading : 6                                  (Other):  2
NA's             :73   NA's      : 3                                  NA's    :107
```

Figure 3: Database summary.

Analysing the database (see Fig. 3), it can be observed that most of the variables have NA's or omitted values. For this reason, it was decided to perform the advanced statistical analysis with FireOrigin1, FireOrigin2 and FireCause, since all possible data are available.

Once the variables are selected to start the analysis, they are defined as active variables in the MCA First, the accumulated inertia (variance explained by the three variables) is calculated for all axes, where there are 21 dimensions. Fig. 4 shows the 10 dimensions with the most inertia only the first 10, which account for 90.2% of the total inertia.

Based on the above information, the variables in the first 4 dimensions will be projected, since 67.3% of the variance is explained. The other dimensions, individually speaking, contribute very little information (<6%) to the database. For this reason, it was decided to interpret only the first four dimensions.

In order to analyse each of the dimensions, two steps will be carried out: first, project the three selected variables to dimensions 1 and 2 and second, the same for dimensions 3 and 4, where the cumulative inertia in dimensions 1 and 2 is 43.3%, and in dimensions 3 and 4 it is 24%.

First, the three variables will be projected in the 1 and 2 dimensions. The projections can be seen in Fig. 5.

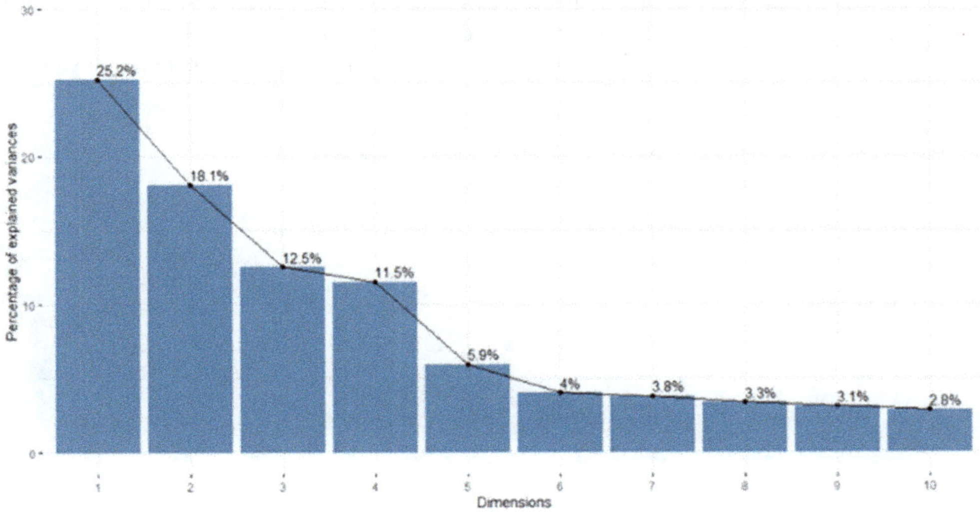

Figure 4: Inertia of the first 10 dimensions (90.2% of variance explained).

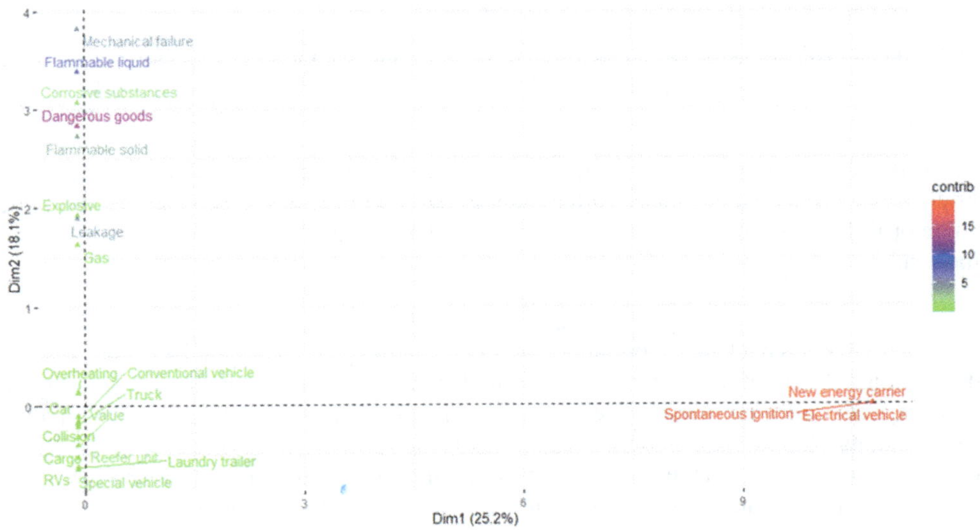

Figure 5: Projections of variables in the first two main axes (43.3%), according to the contributions of their categories.

Fig. 5 shows the categories of the variables according to their contribution in the first two axes (dimensions) from red colour (more important) to green (less important). In the first dimension, the combinations of the following categories are clearly highlighted: new energy carrier, spontaneous ignition and electrical vehicle. In the second axis would be the combination of: dangerous goods, flammable liquid, flammable solid and mechanical failure.

Secondly, the projection will be made in dimensions 3 and 4 (see Fig. 6). The Fig. 6 depicts how dimension 3 is a combination between two categories: Special vehicle and RVs. The fourth dimension is a mix of the categories: value, reefer unit and conventional vehicle.

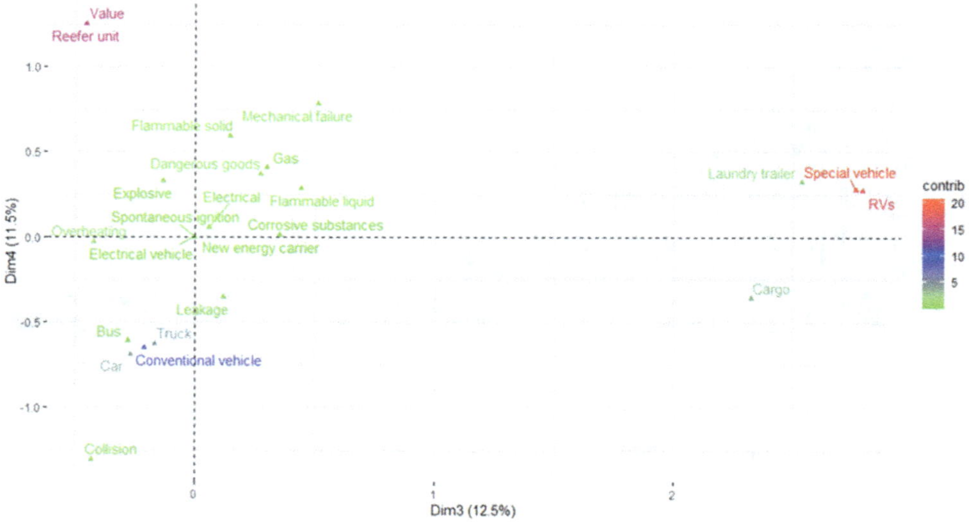

Figure 6: Projections of variables in the three and four main axes (24%), according to the contributions of their categories.

Please note that when FireOrigin1 is reefer unit there are not different options for FireOrigin2, so just value is used in the latter to avoid null values or propagating the same value or RU, for example. It is just a decision on the database design with no relevance on the results.

6 CONCLUSIONS

As a first general conclusion, limitations on the availability of the data (size of the sample, not open-access or not free) lead to a decrease of the quality of the outcomes of the analysis.

From data collection perspective, since there is no standard regarding how information for incidents should be reported, many heterogeneous sets of attributes can be found depending on requirements, functionalities and what is the purpose of the data set. Therefore, a previous time-consumption homogenization of the taxonomies was required. To give some examples:

- Fields describing the ship type usually rely on categories that, again, depend on the source, making it difficult a direct merging.
- Definition of severity include values that depend on a subjective assessment, which may imply a certain level of bias in the results (to be more accurate, in the maritime context, there are clear definitions of the severity level but these definitions are more qualitative than quantitative).

Nothing remarkable from the software engineering point of view, which is agnostic to the abovementioned issues:

- Database design is based on a relational model, which can be extended as needed or implemented with a wide variety of DBMS platforms.
- Using automatisms by means of macros or scripts allows scalability without additional effort in case of updating the starting compilation with new information.

In terms of the MCA data analysis, the results obtained can be summarised as follows:

- In Fig. 5 the first dimension (Dim1 – horizontal axis) shows the combination of the categories new energy carrier, electrical vehicles and spontaneous ignition for the variables FireOrigin1, FireOrigin2 and FireCause. This result represents what happens in the real world, where fires in electrical vehicles are originated in their batteries due to an unexpected and unpredictable thermal runaway.
- Also, in Fig. 5, the second axis shows the combination of: dangerous goods, flammable liquid, flammable solid and mechanical failure. Considering that mechanical failure also includes friction between parts, the combination reveals how important is lashing of dangerous goods during the stowage process to avoid displacement of the units. All incidents of the database belonging to this combination reported ignition as a result of an increase of the temperature of mechanical parts close to flammable goods. Friction was created by means of repetitive movements of the unit due to bad weather conditions.
- In Fig. 6, the third dimension (Dim3 – horizontal axis) shows that the most frequent combination is composed of special vehicle and RVs (recreative vehicles like mobile homes) as FireOrigin1 and FireOrigin2. Into this combination, the most usual fire cause comes from the cargo (the nature of these vehicles drives to a wide range of items that could represent a fire hazard) and, secondly, due to electrical problems, which represent a subset of the previous, including all kind of gadgets that have electricity as their main power supply like stoves, ovens or fans with non-standard installations.
- Also, as it is shown in Dim4 (vertical axis) in Fig. 6 that there is a strong relation between conventional vehicles (FireOrigin1) and both car and truck (FireOrigin2) which makes sense since these two types of cargo units are the most usual being transported in ro-ro ships.
- Finally, for both reefer units and conventional vehicles, the causes are significantly variable, resulting on a point cloud (green colour). Thus, no definitive conclusion can be extracted from the analysed sample. However, as additional conclusion, the most causes are electrical (28%), leakage (23%), overheating (10%) and spontaneous ignition (8%).

ACKNOWLEDGEMENTS

The research presented has been financed as part of the LASH FIRE project (Legislative Assessment for Safety Hazards of Fire and Innovations in Ro-Ro Ship Environment), funded by the European Union's Horizon 2020 research and innovation programme under grant agreement No. 814975. Please visit the website http://lashfire.eu for further information.

REFERENCES

[1] IMO, Sub-committee on ship systems and equipment (SSE 6), 4–8 Mar., 2019. https://www.imo.org/en/MediaCentre/MeetingSummaries/Pages/SSE-6th-session.aspx. Accessed on: 13 Jul. 2022.

[2] LASH FIRE, Deliverable D04.1: Review of accident causes and hazard identification report, May 2020. https://lashfire.eu/. Accessed on: 13 Jul. 2022.

[3] Wikman, J., Evegren, F., Rahm, M., Leroux, J., Breuillard, A., Kjellberg, M., Gustin, L. & Efraimsson, F., Study investigating cost effective measures for reducing the risk from fires on ro-ro passenger ships (FIRESAFE). EMSA, 2016.

[4] LASH FIRE, Deliverable D08.1: Definition and parametrization of critical fire hazards, classification of cargoes, transport units, engines, fuels and vessels and identification methodologies, June 2021. https://lashfire.eu/. Accessed on: 13 Jul. 2022.
[5] Correspondence Group on Casualty Analysis, Casualty statistics and investigation: Report of the Correspondence Group on Casualty Analysis. IMO, Sub-Committee of Flag State Implementation, 2012.
[6] GISIS, Home page. https://webaccounts.imo.org/. Accessed on: 22 Jun. 2020.
[7] LASH FIRE, Home page, 2022. https://lashfire.eu/. Accessed on: 13 Jul. 2022.
[8] ForeSea, Home page. https://foresea.org/. Accessed on: 22 Jun. 2022.
[9] Comisión Permanente de Investigación de Accidentes e Incidentes Marítimos, Investigaciones realizadas por la CIAIM, 2020. https://www.mitma.gob.es/organos-colegiados/ciaim. Accessed on: 20 Apr. 2022.
[10] Hellenic Bureau for Marine Casualties Investigation, Investigation reports, 2020. http://www.hbmci.gov.gr/indexEng.htm. Accessed on: 20 Apr. 2022.
[11] Ministry for Transport, Infrastructure and Capital Projects, Maritime Transport, Maritime Security, 2020. https://www.transport.gov.mt/maritime-40. Accessed on: 20 Apr. 2022.
[12] European Marine Casualty Information Platform. https://portal.emsa.europa.eu/web/emcip. Accessed on: 20 Apr. 2022.
[13] Gehandler, J., Karlsson, P. & Vylund, L., Risks associated with alternative fuels in road tunnels and underground garages, 2017. http://urn.kb.se/resolve?urn=urn:nbn:se:ri:diva-29101.
[14] Härdle, W.K. & Simar, L., *Multivariate Statistical Analysis*, 4th ed., 2015. DOI: 10.1007/978-3-662-45171-7.
[15] Fernandez, S. de la F., Análisis de Correspondencias Santiago de la Fuente Fernández, Fac. Ciencias Económicas y Empresariales, pp. 58–59, 2011. https://www.fuenterrebollo.com/Economicas/ECONOMETRIA/REDUCIR-DIMENSION/CORRESPONDENCIAS/correspondencias.pdf.
[16] Iyer, R.R., Sharma, M. & Saradhi, V., A correspondence analysis framework for author-conference recommendations, pp. 1–49, 2020. http://arxiv.org/abs/2001.02669.

MOSES PROJECT: ENHANCING SHORT SEA SHIPPING WITH AUTOMATED TECHNOLOGIES

MERCEDES DE JUAN, IGNACIO BENITEZ & JORGE MARCOS
Innovation and Port Cluster Development, Fundación Valenciaport, Spain

ABSTRACT

The MOSES Project (AutoMated Vessels and Supply Chain Optimisation for Sustainable Short SEa Shipping) aims to enhance the short sea shipping (SSS) component of the European supply chain. The MOSES project aims to create sustainable SSS feeder services to small ports with an expected 10% modal shift to SSS in designated areas. To attain this target, MOSES proposes a combination of automated/autonomous technologies and supply and demand matching optimisation, like its AutoDock system for containerships in large terminals, which consists of the MOSES Autonomous Tugboat Swarm, supported by MOSES Shore Tugboat Control Station, and the automated docking scheme. Loading times for feeder vessels in large and small ports are expected to decrease by 20% from the operation of MOSES Robotic Container Handling System outfitted on MOSES Innovative Feeder. Supply chain optimisation aims at increasing the attractiveness of SSS. The Innovative Feeder will be designed for a net reduction of greenhouse gas emissions of more than 90%. The MOSES solution expects to attract stakeholders to SSS for a more sustainable EU container supply chain from the environmental, cost, and societal perspectives.

Keywords: autonomous vessels, port optimization, clean technologies, safety, modal transport, SSS.

1 INTRODUCTION

Ports play a decisive role in the EU's external and internal trade. Although ports and especially deep-sea shipping (DSS) ports are integral nodes within multimodal logistic flows, short sea shipping (SSS) and inland waterways are not so well integrated. As the European Union (EU) has aimed for a modal shift from road transport to other more sustainable transport modes, such as rail or waterborne transport, SSS has gained more attention. After liquid bulk, containers have been the dominant type of cargo in EU SSS. The Mediterranean is one of Europe's major sea basins. The perspectives of container shipping in the western Mediterranean side reveals a trend of expansion of major infrastructure in the main ports while in the eastern side of the Mediterranean Sea, Black Sea and Suez Canal offer a significant opportunity in the container transportation to eastern European trade areas. MOSES addresses these challenges in a collaborative manner. This paper presents the main outcomes of this project, developed by a consortium of European companies and research centres, such as MARIN, Trelleborg, CIRCLE, Fundación Valenciaport, McGregor, coordinated by the National Technical University of Athens (https://moses-h2020.eu/).

The aim of the MOSES project is to enhance the SSS component of the European supply chain by addressing the vulnerabilities and strains related to the operation of large containerships. A two-fold strategy has been followed from the beginning, in order to reduce the total time to berth for TEN-T Hub Ports and to stimulate the use of SSS feeder services to small ports that have limited or no infrastructure. Fig. 1 summarises the MOSES concept.

2 IDENTIFICATION OF MARKET OPPORTUNITIES AND BUSINESS CASES

In the context of the MOSES project, a two-fold strategy to reduce the total time to berth for TEN-T hub ports, and also to stimulate the use of SSS feeder services to small ports that have limited or no infrastructure, have been followed in this assessment. In this respect, a typology

WIT Transactions on The Built Environment, Vol 212, © 2022 WIT Press
www.witpress.com, ISSN 1743-3509 (on-line)
doi:10.2495/UMT220151

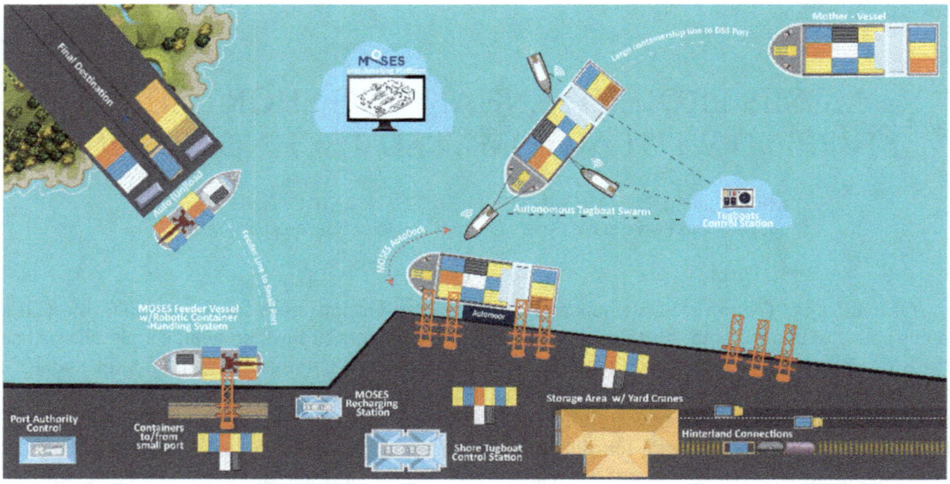

Figure 1: MOSES concept.

of the ports was suggested for the Innovative Feeder vessel and Robotic Container Handling System innovations.

Furthermore, the market opportunities have been exploited for the MOSES innovations in the context of the SSS part of the container supply chain and develop the MOSES Business Cases: western Med-Spain and eastern Med-Greece, in view of the MOSES Transferability Business Case. The focus was on a preliminary analysis to evaluate the viability for container feeder lines, with some innovations in the feeder vessel, versus the current transport system for the case of the port of Piraeus, and also for the case of the port of Valencia. As regards the port of Piraeus, the scope of work concerns a potential feeder line between the port of Piraeus, Mykonos, and some nearby Greek islands. In the current situation, all the general cargo traffic between Piraeus and Mykonos and nearby islands is handled by Ro-Ro or Ro-pax vessels. On the other case, for the port of Valencia, the scope of the work concerns a potential feeder line between three ports managed by the Valencia Port Authority: Valencia, Sagunto, and Gandía. For our case, the containers that reach Sagunto and Gandía hinterlands are loaded/unloaded in Valencia port.

For the case of Piraeus, it concluded that nowadays, the whole general cargo traffic between Mykonos and nearby islands is Ro-Ro traffic, composed of trucks and trailers. Therefore, to develop a container feeder line, the cargo should not only be transferred from the Ro-Ro logistic chain to the Lo-Lo logistic chain but should also be consolidated/unconsolidated in a different way, which will be a significant added difficulty. The expected demand for the container Greek feeder is shown in Table 1.

For the case of Valencia, the analysis showed that despite the longer land distances to the Sagunto and Gandía hinterlands, the shortage or absence of container lines in Sagunto and Gandía port has as a result that most of the traffic is handled through the port of Valencia and carried to the hinterland by truck or rail depending on the case. To evaluate the possibility of developing a container feeder line that links the three ports managed by APV (Valencia, Sagunto, and Gandía), it has been proposed a feeder service with a frequency of three weekly services, with geared ships. And it has been observed that the feeder line would be competitive around Sagunto and Gandía, where the expected container demand for this line is shown in Table 2.

Table 1: Container potential traffic of the feeder line Piraeus port and the islands of Mykonos and nearby islands.

Port	Full containers			Empty containers (TEUs)	Feeder line demand (TEUs)
	Load (TEUs)	Unload (TEUs)	Total (TEUs)		
Paros	730	999	1,729	269	1,998
Mykonos	**2,353**	**34**	**2,387**	**2,319**	**4,706**
Naxos	870	149	1,019	721	1,740
Kea	647	1	648	646	1,294
Syros	569	102	671	467	1,138
Tinos	461	65	526	396	922
Total	**5,630**	**1,350**	**6,980**	**4,818**	**11,798**

Table 2: Container demand of the feeder, by ports and cases (TEUs/year).

Port	Region	Demand captured by feeder line (TEUs/year)		
		Pessimistic scenario	Base scenario	Optimistic scenario
Sagunto	Castellón	13,697	18,263	22,829
	Valencia	38,832	51,776	64,719
	Total Sagunto area	**52,529**	**70,039**	**87,548**
Gandía	Valencia	25,203	33,604	42,006
	Alicante	1,982	2,643	3,303
	Total Gandía area	**27,185**	**36,247**	**45,309**
Total per year		79,714	106,285	132,857
Average per week		1,532	2,043	2,554
Average per service (assuming three weekly services)		511	681	851

3 MOSES INNOVATIVE FEEDERS

The MOSES project ambition is to design innovative feeder vessels. The innovations for the vessels were concentrated on two different directions, in which the innovations for the maritime industry are being developed and demanded by society. The first one was the ambition to reduce or eliminate harmful emissions by designing environmentally friendly vessels. The second one was the ambition to design a highly autonomous feeder vessel, which is able to sail large part of its route without human intervention. Main drive here was to reduce the number of accidents due to human errors followed by a reduction in cost.

The design of both zero emission and autonomous vessels of this size is beyond the current state of the art of technology. Especially for seagoing vessels the steps that are made in this field are very small. Up to now most zero emission and/or autonomous vessels are operated in a small and coastal area or at inland waters. The design of larger seagoing vessels with a capacity of 100 or more than 600 TEU (twenty-foot equivalent unit) and an increased autonomy range, has not been done before. In addition, the vessels were equipped with an autonomous robotic crane.

Concept designs were made using information from a Greek and Spanish business case in combination with the concept of operation and requirements.

In the design of the vessels, partly autonomous operation of the vessels has been considered. This was done in the selection of the equipment but also considering the effects autonomous operations might have on the potential safety of the vessel.

3.1 Zero-emissions concept

With the first ship parameters, the route and cargo volumes, a so called "trip simulation" was done. In this simulation, the sailing and logistic operations are calculated for a period of two years. From the results of these simulations, it was concluded that the presented logistic ambitions could be fulfilled with a smaller vessel, which lead to the introduction of a second Greek case for which the design process was repeated.

The design studies for the Greek case showed that it is technically possible to aim for three different power solutions.

(1) A hybrid powered solution, where a methanol fuelled ICE in combination with a battery, which enables partly zero emission operation of the feeder vessel.
(2) A hydrogen fuel cell option is also an option space-wise but comes with many complications. Particular attention in terms of safety, maintenance and bunkering. Bunkering of hydrogen should be feasible time-wise, as the ship's schedule is typically not very busy, but compressed hydrogen bunkering could still take many hours and requires five truckloads of hydrogen, with support of a land based compressor to reach 700 bar. As an alternative, a large shore-based facility is needed to fuel hydrogen.
(3) A full battery-electric alternative seems disproportional given the large amount of battery storage needed at cost of both available cargo space and a reduced payload.

From these three options, a hybrid powered solution is the most appropriate, making use of shore power and charging during sailing, providing sufficient battery capacity to operate the vessel in the ports of the Greek islands. At open sea, the vessel will be using methanol as a fuel for sailing and charging of the battery.

The operational cost of the hybrid methanol solution (1) is found to be about 15% more than those for a conventional feeder vessel. This is attributed to the higher price of the selected energy carriers for the innovative feeders. The cheapest solution is a vessel with a capacity close to the requested demand per trip rather than an oversized one.

The design study for the Spanish case showed that three power options are feasible.

(1) A battery-electric option is technically realistic, despite that it also requires a rather large battery compared to existing electric newbuilding's, which are much smaller compared to this vessel. In comparison to the total available volume and weight, the battery option seems to fit well.
(2) A methanol hybrid powered solution is more compact, but still features a relatively large battery (6.5 MWh compared to 29 MWh for full electric).
(3) Liquefied hydrogen may be an option space-wise, but compressed hydrogen would require additional deck space which is currently not included in the design. In any case, when it comes to hydrogen, there are many complications to consider. Particular attention is needed in terms of safety, maintenance and bunkering. Due to the limited free time in the schedule, bunkering at a designated area (as required for hydrogen) may be completely unfeasible. Instead, battery charging can be done when moored below a crane, facilitating sufficient time for batteries to be charged.

For the Spanish case both a methanol hybrid powered solution and a battery electric solution were found to be feasible from a technical and operational point of view. The operational cost of the methanol hybrid solution is about 10% lower than those of a conventional design. The cost reduction is derived via the increased autonomy and manoeuvrability of the vessel, which results into lower cost for pilotage and tug assistance.

The PHA for the innovative feeder concepts has resulted in a non-exhaustive list of 12 hazards that are the sources of specific hazardous events. The identification was a result of brainstorming sessions and structured approach to rating the frequency and consequence severity of each event. The identified events relate to the following system components, focusing on innovative aspects of the feeder:

- Propulsion machinery;
- Fuel/energy storage system;
- Cargo space;
- Superstructures;
- Hull.

Most relevant identified hazards are listed in Table 3 and have included technical failures, human factor issues, properties of the energy source used, and issues related to the limits of the operational envelope. In addition, the hazards relate to the following operational phases of the feeder: sailing, cargo operations, and hazards that can arise during all operational phases.

The concept designs of the innovative feeder vessels prove to be feasible for a zero emission and an autonomous ready design feeder vessel for routes that currently are not being served. Thus, the concepts are not only new, but the various evaluations of the selected hybrid concepts show that the realisation of these concepts is feasible and beyond current practise.

Finally, it must be noted that the future costs of the innovative vessels will be lower than those of the conventional ones once a tax related with CO_2 emissions is introduced.

3.2 The autonomous crane

To increase the possibilities for expanding container supply chain to ports where the infrastructure is not supporting lifting containers from/to vessel, the feeder vessel will be outfitted with a robotic container-handling system that is autonomous, i.e., self-sufficient in terms of (un)loading containerized cargo.

Because the on-board robotic crane is able to operate autonomously, i.e. does not need an onsite or on-board crane driver, it is a key innovation for enabling a short-sea container supply chain completely independent from small port infrastructure. In short, the robotic container handling system is designed as a fully self-supporting system that does not need any local help except a quay for berthing and for placing the containers, while maintaining safety of operations.

The MOSES scope is:

- Develop and demonstrate a Robotic Container Handling System (RCHS) capable of handling standard containers.
- Shore control interfaces to landside infrastructure and vessel loading computer.
- Creating a 3D-world model, including sensor suite, for the robot container handling system.
- Virtual reality assisted remote supervisory control of the robotic container handling system.

Table 3: Summary of top hazardous events.

(System comp.) Hazard (ID)	Hazardous event	Consequence	Risk mitigation measures
(Propulsion machinery) System complexity	A technical failure of the hybrid propulsion installation cannot be handled by the crew on board	• The need for specialised support leads to unplanned downtime • Disruption of operation while sailing may lead to injuries/fatalities • Personnel on board unable to manage complex incident	System design (e.g. operator's assistance, autonomous features) and manning need to account for equivalent or higher safety level at engineering compared to baseline, reduction of human error compared to baseline, and optimal number of crew on board for handling systems complexity (issues to be addressed in T3.2)
(Propulsion machinery) Extreme weather/ technical failure	Failure of the generator system when it cannot cope with the load variations in extreme weather conditions	• The energy conversion system will not follow the dynamic loading required by the system. DC grid voltage may drop resulting in a loss of available propulsion power • Possible need to abort operation and subsequent failure to meet demand	• More diesel fuel will be injected to increase the engines dynamics • Alternatively, the battery system should be designed to cope with the dynamic loading as it has a much faster response time. The generator should then only need to run at the average load, and the batteries will provide power in the peaks, and will be recharged in the "lows" • Assessment of system power capacity (incl. redundancy) and power/battery management systems' capacity to handle extreme weather events
(Cargo space) Crane operation	Position of the container crane on board impedes operation of port cranes	Slower cargo handling when using port cranes	• Optimisation of the crane's position on the main deck • Detailed risk assessment to identify required safety systems/barriers during cargo operations • Further considerations in the detailed design of the crane to include limits switches and emergency stop, according to Class rules

Table 3: Continued.

(System comp.) Hazard (ID)	**Hazardous event**	**Consequence**	**Risk mitigation measures**
(Cargo space) Situation awareness	Relative longitudinal position of accommodation and crane limits visual monitoring of the cargo space from the bridge	Fire in cargo space or cargo shift/loss is not detected on time due to failure in manual supervision	• 360-degree visibility from bridge • Equip the cargo space with sensors for early detection with less dependence on visual monitoring • (DNV-RU-SHIP Pt.6 Ch.5 Sec.2), for open top container holds the mechanical ventilation is interpreted to be required only for the lower part of the cargo hold for which purpose ducting is required
(Hull) Extreme weather conditions	Water accumulates in cargo hold in harsh weather conditions due to open top design	Stability degradation and damage to cargo	An analysis based on model tests as well as intact and damage stability assessment following the procedure and the requirements described in MSC/Circular.608/Rev.1 respectively shall be carried out. The additional class notation **Hatchcoverless** includes requirements related to this risk, including for example strength requirements in the intact flooded condition (DNV-RU-SHIP Pt.6 Ch.5)

4 AUTODOCK SYSTEM

Due to increased automation-related complexity, there are challenges for autonomous tugs regarding model uncertainty/error, dynamic forces/moments and changes in hydrodynamic coefficients), also present in auto-pilot systems. MOSES relies on systematic robust control design methods, Robust Model Predictive Control and Robust Optimal (H-infinity) Control, successful in many areas, including the control of surface vessels and underwater vehicles Another issue is the complex communication/control among the autonomous tugboat swarm units that increases computational overheads and makes monitoring and fault detection difficult, addressed by applying an appropriate leader–follower cooperative control structure. MOSES provides scalable autonomous operations, in terms of the required number of autonomous tugboats for each task, which make the solution appropriate for ports of various sizes. Finally, MOSES addressed the necessary integration of the swarm of tugboats' intelligent control system with the AutoMoor back-end.

Development of the communication interface between all vessels (tugs and ocean-going vessels) and the automated mooring systems is also a significant challenge.

Currently, AutoMoor requires human input to acknowledge in terms of confirming and acknowledge that the vessel approaching the berth has been correctly identified via automatic identification system (AIS), allowing the mooring operation to commence and providing response to the system warnings.

The advancements of MOSES innovative AutoMoor units include:

1. Developing and converting all of the human confirmation inputs into full autonomy between the tug vessels, ocean-going vessel and the AutoMoor system.
2. Development of an automated mooring product that can moor shallow freeboard and lightly hulled vessels.

4.1 Tugs swarm

Existing tugboat operations use manually operated conventional tugboats that assist large vessels for manoeuvring in port. To cross over from the current system to an autonomous tugboat swarm system, various milestones need to be accomplished. The milestone of creating and developing a virtual environment, which will enable the training of an autonomous tugboat swarm is described in Fig. 2.

This virtual environment is constituted by various objects that represent real-life components, namely water mass, port, containership, and tugboats. To achieve the best possible representation and producing reliable results, the training environment components are introduced with specific characteristics and properties, derived from extensive analyses and simulations. In this way is ensured that the virtual training environment will utilize realistic forces during training of the AI algorithms. For that purpose, a series of hydrodynamic simulations is provided, offering valuable parameters regarding the motion and response analysis of the ship objects. In addition, a finite element structural analysis is presented, which assesses the structural response and the limits of the components in various operations. An example of the simulations conducted is provided in Figs 3 and 4.

The results from this study indicate the maximum allowable approach speed and contact force for the tugboats. Lastly, a validation was processed by defining two categories of test scenarios evaluating respectively the individual behaviour of objects and the interaction between them in the training environment. The results of the first category showcased a good agreement whereas the results of the second a larger disparity led to the conclusion that some adjustments need to be considered.

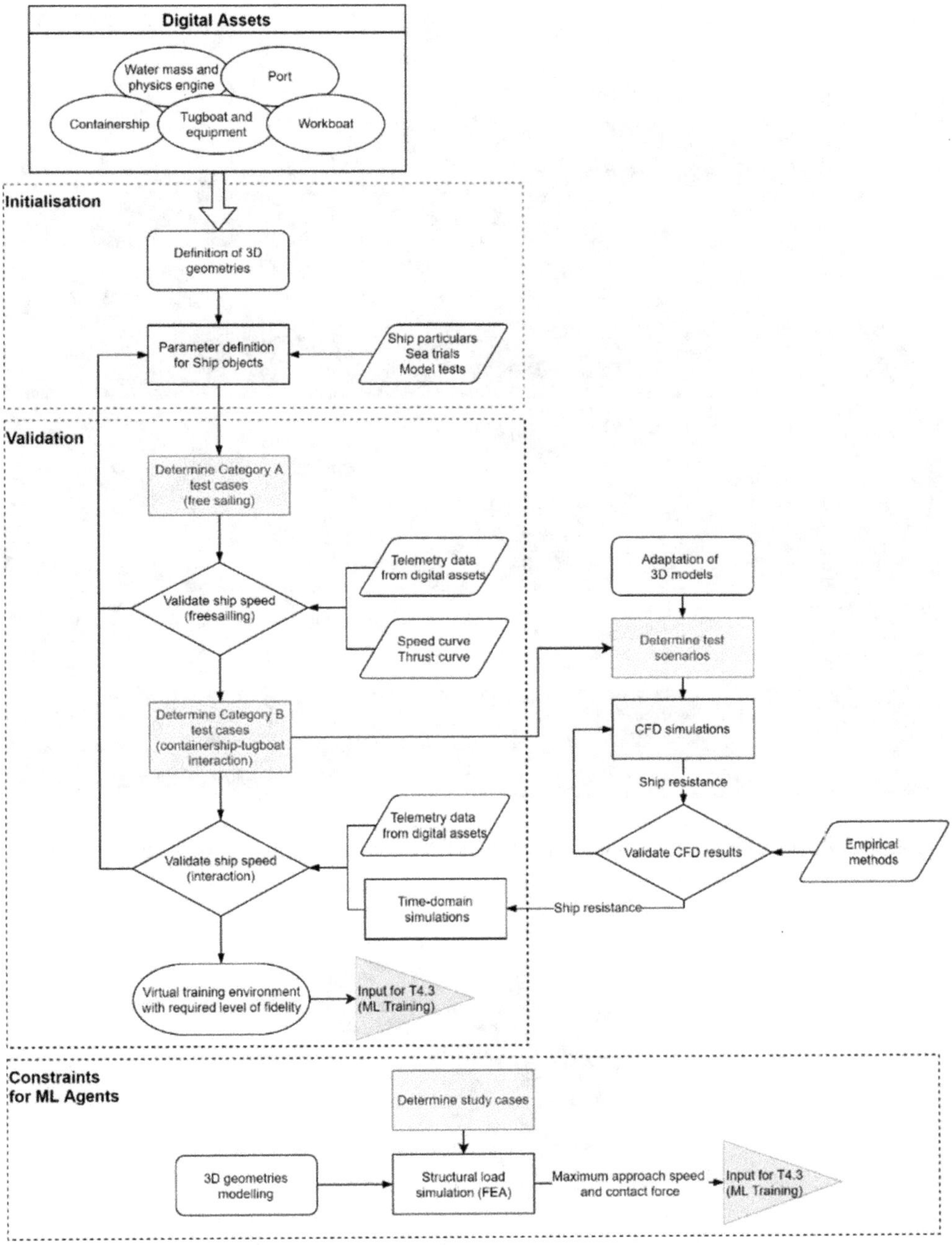

Figure 2: Methodological framework for the development and validation of the virtual training environment.

Figure 3: Various views of transverse pushing of the containership by two tugboats in Unity.

Figure 4: MOSES innovative AutoMoor unit.

4.2 MOSES innovative AutoMoor units

The goal of the MOSES project is to push forward innovation and improvements in the design and operation of ports. To achieve this, it is key to address the current challenges the shipping industry faces. Within the scope of automated mooring systems, integration within a fully autonomous port system will be considered, as well as reducing the scale of the AutoMoor unit and its vessel connection system and how the automated mooring control system interfaces with the other port systems. These will allow safe and efficient implementation of automated moorings within the port operations. The updated version of the AutoMoor unit is shown in Fig. 4.

For Trelleborg's future innovations and development of the MOSES project and system platform itself, two areas for innovation have been highlighted. The first being the implementation of rubber damping elements to allow and control surge motion of a connected vessel. The key aspects for improvement are the simplification of surge motion system compared to current hydraulic system used and a reduction in overall costs and energy intensity of the contained parts. The second is the implementation of energy harvesting systems within the scope of automated mooring actuation systems to minimize the environmental impact of the equipment during operation. As there is natural movement of vessels at berth due to environmental factors, and the typical mass of vessels is in the order of thousands of tonnes, even harvesting a small fraction of this energy could offset the net energy consumption of the system. With these innovations, Trelleborg aims to improve the systems performance, reliability, cost and safety and increase port operators access to automated mooring systems.

Based on the optimization of the current automated systems and the scope of new innovations, Trelleborg aims to deliver a key component of a fully autonomous port system.

5 CONCLUSIONS

MOSES is an ambitious project that bares significant innovation potential in the context of European SSS uptake. Its innovation potential covers both vessel design aspects as well as software tools and accompanying governance models to improve related logistics processes.

MOSES innovations will be evaluated with sustainability criteria (environmental impact, safety assessment, cost–benefit analysis) and validated by conducting model tests in test basins and small-scale pilots in relevant testing environments. At the time this paper was written, testing has not yet begun, so the conclusions that can be drawn are very generic. Although the results have only been validated theoretically, it can be concluded that MOSES proves the viability of opening new SSS routes that allows to decongest large ports, while connecting areas relatively isolated.

Namely, the MOSES feeder concept will allow to operate in remote and isolated areas without any specific port facilities. However, the operational cost are 10–15% higher compared with a feeder powered by fossil fuels.

The Autodock system allow safe and efficient implementation of automated manoeuvring and mooring within the port operations. The key aspects for improvement are the simplification of surge motion system and a reduction in overall costs and energy intensity of the contained parts. In addition, collaborative performance of the swarm has been technically proved. With these innovations, MOSES aims to improve the systems performance, reliability, cost and safety and increase port operators access to automated manoeuvring systems.

ACKNOWLEDGEMENT

This research has been conducted as part of MOSES project, which has received funding from the European Union's Horizon 2020 Research and Innovation Programme under Grant agreement No. 861678. Content reflects only the authors' view and the agency is not responsible for any use that may be made of the information it contains.

REFERENCES

[1] ANSYS Fluent Theory Guide, Ch. 4.4.3 ANSYS, R2, p. 61, 2021.
[2] Holtrop, J. & Mennen, G.G.J., An approximate power prediction method. NSMB Paper 689, 1982.
[3] Birk, L., Resistance estimates: Guldhammer and Harvald's method. *Fundamentals of Ship Hydrodynamics: Fluid Mechanics, Ship Resistance and Propulsion*, Wiley, 2019. DOI: 10.1002/9781119191575.ch31.
[4] Wageningen B series. Standard series of propeller geometry. https://www.marin.nl/en/jips/cd-propeller-original.
[5] Blendermann, W., Wind loading of ships, collected data from wind tunnel tests in uniform flow. Institut fur Schiffbau der Universitat Hamburg, 1996.
[6] DNV-GL: Assessment of selected alternative fuels and technologies, June 2018.
[7] IMO MSC.1/Circ.1621, 7 December: Interim guidelines for the safety of ships using methyl/ethyl alcohol as a fuel.
[8] Van Hoecke, L., Laffineur, L., Campe, R., Perreault, P., Verbruggen, S.W. & Lenaerts, S., Challenges in the use of hydrogen for maritime applications. *Energy Environ. Sci.*, **14**, pp. 815–843, 2021.
[9] MSC/Circular.608/Rev.1 Interim Guidelines for Open-top Containerships. Adopted on 5 Jul. 1994.
[10] MOSES: Automated Vessels and Supply Chain Optimisation for Sustainable Short Sea Shipping. HORIZON2020 EU funded project under grant agreement no. 861678. Jul. 2020–Jun. 2023.

M/V EVER GIVEN: LEGAL ASSESSMENT OF THE CAUSES AND CONSEQUENCES OF THE ACCIDENT

MICHAEL BOVIATSIS
Department of Maritime Studies, University of Piraeus, Greece

ABSTRACT

The grounding of *M/V Ever Given* in the Suez Canal for six days led to an unprecedented perturbation to the supply chain of the maritime sector. Specifically, the sudden closure of the canal resulted in the delay of hundreds of vessels and subsequently to enormous claims and even, in some cases, loss of cargo. The *M/V Ever Given* was later impounded by Egyptian authorities for an initial claim of $916 million and was released after three months when the Egyptian government and the managing company reached an agreement for compensation. This paper aims to assess the most prominent causes of the accident, locate the causa proxima that directly led to the accident and evaluate the impact of the relevant legislation under which the vessel was operated. Initially, the material facts regarding the state of the vessel, the canal, and external factors will be explored, and each action that possibly led to the accident will be analysed. The outcome of this analysis will provide the causa proxima of the accident, and then an evaluation will be made regarding the liabilities of each stakeholder. The analysis will continue with evaluating the impact of the national and international legal provisions that were in effect during the crossing of *M/V Ever Given* from the Suez Canal and with the assessment of a possible limitation of liability of the most liable party. The analysis will be based on the legal doctrine, evaluating relevant legislation and previous court decisions. Finally, in conclusion, measures will be proposed to be implemented, aiming to prevent the repetition of such accidents in the future.
Keywords: SCA, grounding, Hague–Visby Rules, COLREGS, causa proxima.

1 INTRODUCTION

Suez Canal's total closure due to the *M/V Ever Given's* grounding set a historical precedent. Not only has it caused an economic uproar, but it has also erupted in multifaceted legal implications. Therefore, before carrying out a legal assessment of the rights arising from the accident, the priority is to find out the main cause that inevitably led to the accident. This is because it is the prerequisite for defining and assigning legal responsibilities and rights in order to clarify which articles of Hague–Visby Rules can be applied since, with them as a legal basis, the result of our inquiry will be judged [1].

The Suez Canal's closure, caused by the grounding on 23 March 2021 of the mega container ship *M/V Ever Given*, highlighted in the harshest way the uneven development of sea canals in relation to the ship repair industry and the enormous legal challenges of such incidents. In the seven days of non-operation of the canal, just under 400 ships with their cargoes were waiting to pass through the canal, creating multifaceted maritime claims and economic consequences. Below the causa proxima will be examined between the accident and the event preceding the accident along with: the respective attributions of responsibility, the legal consequences for the ship owner and the factors to which a ship owner should pay attention by studying the case study of *M/V Ever Given* [2].

As for the economic liabilities, it is estimated that, during the blockage of the canal, 12% of world trade or 30% of international container capacity was prevented from passing through or forced into a deviation solution from South Africa and the Cape of Good Hope [3]. Products, from cattle to crude oil and containerized cargo, worth almost USD 67.2 billion, remained in the canal, some of which were severely damaged [3]. Lost profits for the Suez Canal Management Authority are estimated to have been $13–$15 million per day.

WIT Transactions on The Built Environment, Vol 212, © 2022 WIT Press
www.witpress.com, ISSN 1743-3509 (on-line)
doi:10.2495/UMT220161

According to the Freightos Baltic Index, "freight prices on the China/East Asia line to the Mediterranean have increased by 500% in March 2021" [4].

2 THE WARRING PARTIES AND DESCRIPTION OF THE ACCIDENT

2.1 Description of the Suez Ship and Canal

The *M/V Ever Given* flag of Panama is a container ship and is one of the largest in the world with a length of 399.94 m, a width of 58.8 m and a draft of up to 16 m, while its carrying capacity can reach 20,124 TEU where it has time chartered and still manages ten similar such ships [5].

The Suez Canal is an artificial waterway connecting the Red Sea with the Mediterranean and separating Africa from Asia, offering ships a direct route between the North Atlantic and the northern Indian Ocean through the Mediterranean and Red Seas, avoiding the passage of Africa (travel reduction of 5,500 nautical miles (nmi)) and the dangerous passage of the Cape of Good Hope. Its construction began in 1859, and the inauguration of the canal took place on 17 November 1869. The canal, which has a length of 120.11 nmi, is operated and maintained by the State Authority of the Suez Canal (SCA) of Egypt and allows ships with a maximum width of 77.5 m and 20.1 m draft to cross it. In 2020, more than 18,500 ships crossed the canal (51.5 per day on average). Also, in 2016 there was an extension of the canal, with a detour of 22 nmi to reduce transit time and almost double ship tonnage from 49 to 97 ships per day [6].

Concerning the rules of navigation, in accordance with Article 6 of the Rules for Navigation in the Suez Canal, it is submissive to use a pilot throughout the passage of the canal and for all ships.

However, in accordance with Article 4, the master, the owner of the ship and the charterer shall be liable for any damage or accident caused during the transit of Suez unless evidenced that this damage was not done intentionally or through error or negligence. This liability also applies to environmental damage, if any. As understood, the navigator is mandatory with an exclusively advisory role, signed by the master before entering the canal [7].

2.2 The chronicle of the accident

On 23 March 2021, the *M/V Ever Given* was sailing from Tanjung Pelepas to Marseilles through the Suez Canal, which it had crossed 22 times since its construction. On the morning of that day, the ship ran aground with the bow but blocked the entire width of the canal, interrupting traffic from both directions. According to the first estimates, powerful winds were blowing that exceeded 40 knots. In combination with the volume of the ship, they made it deviate from its course and unable to perform manipulations [1].

Immediately after the grounding, the channel authority banned traffic, which resulted in the congestion of at least 369 ships (which accounted for about 16.9 million DWT) until 29 March 2021, when they managed to release the ship and allow the passage of the canal again. The ship, after detachment, sailed towards the salt lake and, on 13 April, was seized by the Egyptian authorities at the request of the SCA, where they were seeking compensation of over $916 million, including $300 million for the ship's detachment and $300 million for "reputational damage". On 7 July 2021 Egypt released the vessel following an agreement between the authorities and the ship-owning company for compensation [3].

3 POSSIBLE RESPONSIBILITIES FOR GROUNDING

3.1 Definition of the canal and status of Suez

In order to understand where the blame lies, the type of water in which the ship was moving must be assessed. A canal is defined as any artificial – terrestrial – project often of great width and length in order to connect seas, lakes and rivers. Such canals, mainly used for waterborne transport, are also characterised as "floating canals" or navigational canals and are created for purposes, primarily commercial, by joining different All kinds of waterways can be navigated [8].

According to international law, "canals are artificial routes established within the territory of the state and subject to its sovereignty", so freedom of navigation depends on the consent of that state. On the contrary, canals that are a passage of major sea routes and are of great interest in navigation in general, both during the war and during peace, such as the Suez and Panama Canal, are in an internationalised legal status where free passage applies [9].

Despite what has been mentioned above, the SCA allows passage through the Suez Canal to ships of all nations subject to compliance with the conditions set out in the navigation rules and explicitly states that all petitions and circulars will be issued by the SCA will be an integral part of these rules. Ships must also comply with the provisions of the SOLAS, MARPOL 73/78 International Regulations, as well as the requirements of the International Regulations for the Prevention of Collisions at Sea (COLREGS) and all laws, orders and regulations issued by the Egyptian government. Moreover, simply that they use the water of the canal, the masters and shipowners undertake to accept all the terms of the present and only sailing rules, by which they declare that they know and comply with these terms in all respects, comply with any requirement made for the purpose of their proper execution and adherence to the private Code of Signals of the SCA [10].

Finally, the SCA reserves the right to refuse access to the canal or to order the towing or escorting of ships considered dangerous for navigation in the canal. Pilotage is mandatory for all vessels unless the SCA approves it. The masters are held solely responsible for all damages or accidents resulting from the navigation or handling of their ships, for which, throughout the passage, they have exclusive sovereignty [7].

3.2 Causa proxima

In order to find the causes of the incident, three very important factors will have to be analysed, which probably played a decisive role in the ship's grounding. First, according to testimonies of the pilots, the captain and the crew, at the time of the accident, a sandstorm was underway in the area with powerful winds [11].

It was observed that winds of up to 70 knots prevailed, while due to the sandstorm, the wind was denser than usual as it contained many soil particles and dust. This, combined with the substantial volume of the vessel due to the container, makes it vulnerable to strong winds, making proper steering difficult [11]. It should also be noted that when the wind speed exceeds 10 m/s, the required angle of displacement of the rudder must exceed 35° for the ship to "hear", which leads to the inability to provide adequate control [12].

Another factor that played a role was the speeding of the ship. The SCA accused the captain of speeding, which according to the Egyptian authorities, was the cause of the grounding. The ship did not have the required reaction time, and in combination with the disagreement between the master-navigator who heard from the VDR about the steering wheel of the ship, a possible delay in order to change course may have led to the grounding [13].

It was also found that the ship's speed was increased compared to the usual vessel speed. Finally, the size of the vessel also played its part, as this particular ship is one of the largest ships in the world that sailed in enclosed waters because these ships are cumbersome in terms of manoeuvres and are easily affected by weather conditions (strong winds) [11].

In conclusion, if all three of the above factors are taken into account, it can be stated that the ship's grounding was almost inevitable. The most likely cause of the accident, however, was the strong wind, which was aggravated by a significant increase in the density of the air–dust mixture, resulting in a loss of control and loss of control – subsequent unpredictable behaviour of the vessel. Therefore, the only way to avoid the accident was to predict the wind speed and prohibit the ship from entering the Suez Canal if there was a risk of a sandstorm [12].

Because of the above, the straightness initially weighed on the principle of the canal as it allowed entry into the *M/V Ever Given.* At the same time, poor navigation conditions prevailed for ships of this type. Still, although the ship had a higher speed than it should have been, the traffic of the vessels was monitored from the beginning of the canal and by the navigators on board [13].

However, although while the channel is artificial, the responsibility for whatever happens is held by the master, which is signed before entering the canal. So, the master is obliged to intervene if he saw something wrong with the ship's course or the dangerous weather conditions [7].

4 LEGAL ASPECTS

4.1 Legal assessment of the arguments stated by the parties

This section is important for the subsequent attribution of the limits of the legal rights of each side depending on which of the events is considered to be the causa proxima, that is, the direct cause that inevitably caused the accident, i.e., the direct cause is not simply closer in time to the loss, but is more comparable in effectiveness to bring about the accident, as in the *Leyland Shipping Co Ltd v. Norwich Fire Insurance Society Ltd* case [14]. In particular, at the 101st km of the canal and the narrowest point of this ocean liner, the *M/V Ever Given* lost control, causing the ship to run with its bow on one side of the canal and due to its rotation on the channel to block traffic. According to its characteristics, it is a mega containership, among the ten largest container ships in the world [15].

The guaranteed depth of the Suez Canal is 24 m, and the width at a depth of 11 m reaches 225 m. Therefore, a vessel less than 200 m long should not face difficulties crossing. Furthermore, the turn of the 400 m vessel by only 40° completely clogs the canal, which happened on 23 March 2021 [16]. According to scientific research, two main theories have been formed about the causa proxima of the accident. On the one hand, it is the approach that the accident was caused due to adverse and unpredictable weather conditions [12], with the opposite view blaming the naval incident on the master's mishandling and inadequate decision-making ability, therefore on human error and therefore on the ship's unseaworthiness [17].

Thus, the grounding has been attributed to the narrow canal is some sections, the size of the ship, the strong winds, poor operation and violations of navigation rules. Shortly after the accident, the positions of the involved parts were different and rivalled to some extent.

However, initially, the technical problems of the ship were pointed out as the primary cause of the accident [11].

Suez Canal authorities blame the high-speed and the rudder design for the ground of the MV Ever Given. According to the SCA, when passing through the spot, the ship was moving at a speed of 25 knots when it should have been 8–9 knots. Also, the SCA argued that the Suez Canal has large straight sections in the accident area. Therefore, it requires restrictions on the angle of deviation of the vessel from the axis of the runway, which is determined by the capabilities of the steering device; The rudder was not aligned, and the size of the ship was not suitable for the size of the crossing point. For all the above, the SCA argued that it was the choice of the master of the ship not to let trained pilots of the canal board the vessel to navigate it safely as they were aware of the peculiarities of navigation of the points there. Wind speed at that time was recorded at 40 knots, but the SCA argued that this was not the only reason the ship ran aground [18].

On the other hand, the ship's interest side argued that according to the investigation, the strong winds through the ship off course and then a series of orders from pilots worsened the situation, leading the vessel to crash on both banks of the canal. In addition, the SCA violated the Navigation Rule Section II, Art. No. 11 (navigation) regarding the development of tugs next to any vessel of the size of the *Ever Given* while navigating the channel to avoid direct collision with the channel embankment [19].

In addition, the ship's interest side argued that the real cause of the accident was the loss of control and the subsequent unpredictable behaviour of the vessel due to the extreme and unexpected weather conditions that prevailed at that time. This is because, by the competent authorities, as long as there was a risk of sandstorm [20].

In defence of their position, the ship's interests side briefly argued the following: *Passing through the Suez Canal, the vessel entered the area of a dust storm accompanied by an increase in wind speed and a decrease in visibility. Sandstorms in this region are frequent and no less severe and fraught with consequences than snowfall on the Strait of Kara and the Vilkitsky Strait on the Northern Sea Route* [12].

A dust storm can last from several hours to two or three days. An accident similar to the one in the Suez Canal can occur in any other place and with any boat that has a sufficiently large relative exposed to the wind area (or wind area) moving under the influence of the wind [18].

The effect of the wind on the parameters of the ship's movement directly depends on the ratio of its areas above the water and the underwater body. The presence of deck containers significantly increases the wind area of the vessel. Thus, with an increase in load capacity, wind's effect on movement parameters increases significantly [11], [12].

As noted earlier, while passing through the Suez Canal, the vessel entered the area of a sandstorm that was accompanied by an increase in wind speed. The flow of air during a sandstorm contains a large number of soil particles and dust. Consequently, the density of such a suspension is greater than the density of the usual flow generated in the wind tunnel when determining the aerodynamic characteristics of the body. Thus, the presence of sandy suspension in the air can impair the vessel's manoeuvrability [17] and lead to a loss of control even at a safe wind speed of 9 m/s, as identified in one of the SCA's main arguments [12], [21].

4.2 Possible liability for delays and damages of goods: Cargo claimants

A large part of the legal claims rests with the owners of the cargo owners that were transported with the *M/V Ever Given*, otherwise with the bill of lading holders. The mandatory legal framework is the application of the Hague–Visby Rules, where the latter are mentioned as terms in the unique contract between the carrier (*M/V Ever Given*) and the shipper (cargo owner), which is the bill of lading – as in *The Ardennes* case [22]. According to Art I r. (1)(g). Article X of the HVR states these are mandatory to apply to bills of lading. Also, HVRs are applicable in this case because, at the time of the accident, the cargo was "under the ship's tackle", as in the *Pyrene v. Scindia* case [23].

According to Article III r.6 HVR, the owners of the cargoes have one year, starting from the day on which the products were delivered, as in *The Sonia* case [24], and during which they can take legal action against the carrier. In *The Aries* case [25], the cargo owners' claim had ceased to exist under Art. III r.6 and could in no way be entered into court proceedings by the defence or set-off. In addition, the cargo owners do not have the right to apply to the court to extend this deadline [22]. A robust legal basis, in accordance with the provisions of the immediately preceding chapter, for the owners of cargoes located on the *M/V Ever Given* is Article III rule 1 HVR, in which the carrier is obliged to exercise due diligence before and at the start of the trip, as in *Maxine Footwear Co Ltd* case [26], to keep the ship in a seaworthy condition, where the seaworthiness includes the physical condition of the vessel and the level of manning and management of the vessel.

An essential element for these cargo owners is this obligation of the carrier – owner or charterer of the *M/V Ever Given* that introduces Article III r.1 HVR is a personal duty of care, as in *The Amstelstot* case [27]. Therefore they cannot transfer this obligation, therefore, cannot be waived by this obligation, as in *The Muncaster Castle* [28], by facilitating the work of the owners of consignments. The difficulty for the cargo owners is that according to Article IV r.1, they bear the burden of proving the unworthiness of the *M/V Ever Given* and the causal link between unworthiness, blocking the canal and causing damage to their cargo. In the case of *M/V Ever Given,* there is an objective difficulty in proving the unworthiness as almost all reports of maritime investigation casualties are inaccessible for the retrieval of data, there is a strong opposing view as to the root cause of the accident, and therefore there is serious difficulty in establishing a causal link between the unworthiness and the damage to the cargoes [29].

Cargo owners must remember that if they fail to prove the lack of due diligence, the exceptions of Article IV r.2 HVR will be made available to the shipowner, as in *The Hellenic Dolphin* case [30]. Finally, it is essential for the owners of cargo that any term in the contract of carriage, in this case, on the bills of lading, which is intended to relieve the carrier of the liability or to reduce the liability of the airline, is invalid (Article III r.8 HVR). The judgment in *The El Greco* case is significant, where the court deleted such a contractual clause [31].

4.3 Possible liability for delays and damage to goods: Cargo claimants of deviated vessels and other vessels

First of all, it should be said that eight ships transporting animal stock (live animals) were immobilized due to the accident due to the accident. However, according to Article I(c) HVR, the category of live livestock is excluded from the application of HVR, which does not cover it, so the owners of this load will have to find another legal route to be compensated [3].

Another related issue is that many ships were forced to change course, doing a deviation, using longer sea routes. Many ships have abandoned their original route, diverging to the nearest but longest route – the Cape of Good Hope – to avoid further delays. In the cost of container shipping. Considering that Europe occupies the highest percentage of containers in the world's maritime trade with 23%, the blockage leads to an increase of more than 10% in the cost rate per container. Of course, the above economic losses cannot be claimed through HVR [32].

Deviation options caused a delay in delivering the goods to the cargo owners. It is noted that according to Article IV r.4, the HVR allow deviation from the original course to save life or property or for any "reasonable deviation"; the objective navigation difficulties posed by the canal's blocking must be considered to fall into the category of "reasonable deviation or deviation for reasonable reasons", while also applying the case-law rule laid down in *Reardon Smith Ltd. v. Black Sea and Baltic General Ins. Co.* [33], where it was stipulated that the appropriate route is the direct geographical route (Suez Canal), but if the data are sufficient (total and lasting Suez blockage) to document a practice to follow a particular route (Cape of Good Hope), then this use is the route is not a deviation. The latter judgment clarified the facts of the earlier case law of *Stag Line v. Foscolo* [34], which defined the reasonable deviation that a prudent person would make after taking into account all the relevant circumstances and the contractual obligations and interests of all parties, In this case, the interests of all parties and the surrounding circumstances justified the deviation from the Cape of Good Hope, especially when this deviation had been organized before the ship sailed from the port of departure, so it is considered a *sine qua* non-convincing, as in *The Al Taha* case [35]. This is because, as many ships did after the first day of Suez's blockage. Therefore, it is concluded that cargo owners have no legal basis for claiming damages from their carriers who were late in delivering their goods to them due to deviation.

4.4 Possible liability for delays and damages of the goods: Possibilities of limitation or discharge of liability by the carrier

According to what has been said earlier, regarding the carrier's claims (*M/V Ever Given* interests) that the accident is due to the unexpected and unique weather conditions prevailing in the area, the carrier may find a legal means of defence in Article IV r.2 (c)–(p), i.e. to invoke a danger to the sea (the weather conditions described above) and to exclude its liability towards the owners of cargo [36].

According to recent scientific articles mentioned above, these weather conditions of the sandstorm, in combination with the unprecedented size of the ship, offered, in the opinion of experts, these exceptional risk conditions, so the carrier must be exempted from any damage suffered by the goods transported by it. However, to rely on this exemption, the carrier must prove in any case that the marine hazard that caused loss/damage was unpredictable, so knowledge of the weather is a crucial factor in determining whether the carrier can rely on the exemption or is negligent, For weather conditions. Usually, the latter had to deny him entry to the canal [7], [21].

Of course, if it is proved that the carrier has violated the obligation to maintain the airworthiness of the ship Art III r. one then they cannot use the exceptions to the liability introduced for them by Article IV r.2 (c)–(p), in accordance with the case-law set by *The Kapitan Sakharov* case [37].

5 JUDICIAL DECISIONS ON SIMILAR INCIDENTS

In order to reach a safe conclusion about the vulnerabilities to be attributed to *M/V Ever Given*, similar cases of accidents that occurred in the canal and the logic of court decisions should be considered.

First, the case of *M/V Panamax Alexander* will be analysed. On the evening of 15 July 2018, a convoy of eight ships sailed to the Suez Canal in a north–south direction. At about 17:50, the *M/V Aeneas*, in charge of the convoy, had a problem with the engine and was forced to anchor in the channel because it blocked the traffic. This caused the rest of the vessels further north to take measures to either dock or anchor, initially with the *M/V Sakizaya Kalon* (7th of the series). At the time of the collision, the *Sakizaya Kalon* was anchored on the east bank of the canal, while in front of him was another boat, the *M/V Osios David*, which was also anchored but on the west bank. During the first collision, the two vessels moved forward and collided with *Osios David*, forcing the closure of the main waterway for several hours [38].

As a result, three legal actions were filed. *Sakizaya Kalon*'s owners did the first against *Panamax Alexander*'s owners. The second was by *Osios David*'s owners against *Panamax Alexander*'s owners. The third is by *Osios David* against the owners of *Sakizaya Kalon*. *Panamax Alexander* argued that it could not anchor as it was in an area with undersea cables while accusing the ships *Sakizaya Kalon* and *Osios David* of not notifying it in time that they had anchored. However, the court found *Panamax Alexander* 100% responsible for the accident (which is extremely rare for conflict cases), as Judge Teare found that *Panamax Alexander* had failed to comply with the Collision Avoidance Regulations (COLREG) and, in particular, Rules 5, 7 and 8 [38].

The judge found that "the conflict between *Panamax Alexander* and *Sakizaya Kalon* is due to the failure of *Panamax Alexander* to recognise the risk of collision and the subsequent failure of the *Panamax Alexander* to anchor before the field KM 149 (i.e. before the underwater cables/conductors). Moreover, having failed to anchor before the proper zone as a prudent master would, the *Panamax Alexander* also failed to drop its anchors in time when it could (within the KM 149 field), and this contributed decisively to the collision" [38].

In yet another similar case, the *M/T Tropic Brilliance* had blocked the Suez Canal for three days, causing damage of millions of dollars. The reported ship loaded with 142,000 tons of crude oil was sailing on 6 November 2004. However, during the passage of the canal and according to the torment of the navigator, the rudder was damaged, resulting in the ship running aground on the right bank and preventing the passage of another vessel. After the impact of the *Tropic Brilliance*, dozens of merchant ships were trapped at the two entrances to the canal. One of them, a tanker of Greek interests, was at the anchorage of Port Said, waiting to sail from north to south. The Greek company asked for compensation for the ship's expenses for as long as it was anchored (expenses for the crew's payroll, catering, insurance coverage, etc.) [39].

After the accident, the *Tropic Brilliance* sailed for repair in Eleusis, so it was agreed between the two sides that the civil dispute would be resolved in the Multi-Member Court of First Instance of Piraeus, where Egyptian law and Greek procedure were applied. The court decision came out after ten years. It held *Tropic Brilliance* responsible for the accident, forcing it to pay compensation of €176,000, rejecting his claims that the Greek company would have spent those sums on the operation of the ship anyway [39].

From the above mentioned cases, it can be concluded that the responsibility constantly burdens the ship and the captain, even though Egyptian navigators were at the wheel at the time of the accidents and had control of the vessel. Moreover, the court rulings rejected the directness of the pilots and the canal principle, not counting their contribution to these accidents.

6 INDEMNIFICATION

After the blockage of the canal, questions arose as to the compensation for the delays of the remaining ships and the damage that may have been caused to their cargoes. It is estimated that about 400 ships were "stranded" due to the grounding. All those delayed ships could bring legal action against either the vessel or the Suez Canal, whoever is responsible.

However, it should be considered which ships were already waiting for their turn when the accident occurred, as the passage is artificial and ships that arrived after the incident – and while they had the possibility of an alternative route – it is not reasonable to ask for compensation [40].

Moreover, even for ships that were moored and waiting to cross the canal before the incident, they could only claim compensation for their cargo in the event that they had "Institute Cargo Clauses A" type insurance that also covers the cargo of the vessel [41].

As assessed above in similar incidents, the perpetrator of the accident was called upon to compensate a ship that was late to pass the canal. So paralysed, the incident of the *M/V Ever Given* is very complicated, with many vessels delayed. Moreover, as thousands of different parties are involved (e.g. cargo and ship owners, insurance companies), the settlement process likely takes a long time because it is a complication [42], [43].

Finally, due to the complexity of the incident and a large amount of compensation expected to be demanded, the guilty side will be almost impossible to cope with, while it is almost certain that it will support the existence of force majeure in order to renounce the liability. But even if force majeure is not established, the owners of cargoes or charterers will likely not be compensated because they delayed waiting to pass or chose to make the significantly more time-consuming circumnavigation of Africa [36], [39].

7 CONCLUSIONS

As already mentioned, this paper examined and evaluated the possible responsibilities for delays and damage to the cargo during the grounding of the *M/V Ever Given* in the Suez Canal.

Initially, the legal status of the canal was analysed. It was evidenced that, while the Suez Canal management authority makes it mandatory for a pilot to be present on every ship passing through it, the pilots have a purely advisory role throughout their presence on board. However, they are essentially the ones who are in control of the vessel and are renouncing themselves for any accident that may arise. Also, the masters take on the entire burden of the liability for any accidents that occur, which they sign before entering the canal. At the same time, the beginning of the canal can be refused entry to the channel for someone who feels they do not comply with the regulations.

In addition, the most likely causes of the accident were analysed, and it was evidenced that the most likely cause was the combination of the high speed that the ship had and the powerful winds blowing by it. However, it is evident that the accident could have been avoided with proper prevention measures installed.

Additionally, the possible claims for compensation have been evidenced and how complicated the situation is because of the large number of ships affected by this event has been recognised. Halted ships, waiting their turn to pass through the canal and without the alternative route option, could claim compensation from the culprit and based on previous incidents, could be compensated even in part.

Finally, it can be concluded that the Suez Canal management authority is responsible for the grounding because it permitted the ship to enter the canal, which could have been refused while poor navigational conditions prevailed. However, the most likely scenario is that the captain will be found guilty of the grounding, initially because it was evidenced that in similar accidents, the court did not lay any responsibility to the SCA but also on the pilots, as the master signs that he is responsible for any accident that occurs, and most importantly because even if the canal management authority and also the pilot made wrong actions, the master has the so-called "Override Authority" for cases where he deems that there is a danger to the crew or the ship, so if he judged that the weather conditions were not suitable or the pilot was giving wrong orders at his discretion, he would have to order anchoring the ship within the canal or refusing to enter it until the appropriate conditions are in place.

Thus, while the managing company of the *M/V Ever Given* bear responsibilities along with the vessel's captain, the canal management departments, including the SCA, have not been sufficiently prepared for the growing size of the vessels. In contrast, the facilities and staff of the SCA have not increased proportionally. So, it can be argued that it is a both-to-blame case in which both sides have legal grounds to assert their rights and protect themselves from third-party claims. As for ships forced into deviation, they have several legal advantages over cargo claimers, putting them in the position of privileged to reduce or avoid liability for any damage to the latter's goods.

REFERENCES

[1] Forti, N., d'Afflisio, E., Braca, P., Millefiori, L.M., Willett, P. & Carniel, S., Maritime anomaly detection in a real-world scenario: *Ever Given* grounding in the Suez Canal. *IEEE Transactions on Intelligent Transportation Systems*, **23**(8), pp. 13904–13910, 2021. DOI: 10.1109/TITS.2021.3123890.

[2] Jie, L., Hao, S. & Peng, Z., The scheme of re-floating a grounded vessel and risk analysis based on *MV Ever Given. American Journal of Traffic and Transportation Engineering*, **7**(3), pp. 51–55, 2022.

[3] Clark, A., Stapczynski, S. & Shiryaevskaya, A., The $9.6 billion a day price of a Suez stuck ship. https://gcaptain.com/the-9-6-billion-a-day-price-of-a-suez-stuck-ship/. Accessed on: 16 Sep. 2022.

[4] Freightos Data, Freightos Baltic index (FBX): Global container freight index. https://fbx.freightos.com/. Accessed on: 16 Sep. 2022.

[5] Evergreen Marine Corp., Vessel particulars: Evergreen line. https://www.evergreen-line.com/static/jsp/vessel.jsp. Accessed on: 16 Sep. 2022.

[6] World Shipping Council, The Suez Canal: A vital shortcut for global commerce. https://web.archive.org/web/20180422172437/http:/www.worldshipping.org/pdf/suez-canal-presentation.pdf. Accessed on: 16 Sep. 2022.

[7] Suez Canal Authority, Rules of navigation. https://www.suezcanal.gov.eg/FlipPDFDocuments/Rules%20of%20Navigation.pdf. Accessed on: 16 Sep. 2022.

[8] Churchill, R., Lowe, V. & Sander, A., *The Law of the Sea*. 4th ed., Manchester University Press, 2022. DOI: 10.7765/9781526159038.

[9] Lagoni, R., Canals. *Encyclopedia of Disputes Installment*, **10**, Elsevier, pp. 36–39, 1987. DOI: 10.1016/B978-0-444-86241-9.50018-X.

[10] Nasreldin, A.R.R., Challenges and potential for Egyptian oil spill contingency plan: A comparative study for improving oil spill preparedness: Case study Gulf of Suez and North Sea. Master's thesis, World Maritime University, 2021. https://commons.wmu.se/all_dissertations/1723/?utm_source=commons.wmu.se%2Fall_dissertations%2F1723&utm_medium=PDF&utm_campaign=PDFCoverPages. Accessed on: 16 Sep. 2022.

[11] Popov, A.N., Zelenkov, G.A. & Papulov, D.S., Reconstruction of various navigational scenarios of the "*Ever Given*" ship, including grounding in the Suez Canal using the bridge simulator with up-to-date electronic navigation charts. *Journal of Physics: Conference Series*, **2061**(1), 012114, 2021.

[12] Lebedev, A.O., Lebedeva, M.P. & Butsanets, A.A., Could the accident of "*Ever Given*" have been avoided in the Suez Canal? *Journal of Physics: Conference Series*, **2061**(1), 012127, 2021. DOI: 10.1088/1742-6596/2061/1/012127.

[13] Lee, J.M.Y. & Wong, E.Y.C., Suez Canal blockage: An analysis of legal impact, risks and liabilities to the global supply chain. *MATEC Web of Conferences*, **339**, 2021.
[14] *Leyland Shipping Co Ltd v. Norwich Union Fire Insurance Society Ltd*, AC 350, 365, 1918.
[15] Vessel Finder, *M/V Ever Given*. https://www.vesselfinder.com/vessels/EVER-GIVEN-IMO-9811000-MMSI-353136000. Accessed on: 16 Sep. 2022.
[16] Sánchez, R.J., Perrotti, D.E. & Fort, A.G.P., Looking into the future ten years later: Big full containerships and their arrival to South American ports. *Journal of Shipping and Trade*, **6**(1), pp. 1–20, 2021. DOI: 10.1186/s41072-021-00083-5.
[17] Kamis, A.S., Fuad, A.F.A., Anwar, A.Q. & Monir Hossain, M., A systematic scoping review on ship accidents due to off-track manoeuvring. *WMU Journal of Maritime Affairs*, pp. 1–40, 2022. DOI: 10.1007/s13437-022-00274-2.
[18] Ship & Bunker, Suez authorities blame *Ever Given* for grounding. https://shipandbunker.com/news/world/636420-suez-authorities-blame-high-speed-and-rudder-design-for-ever-given-grounding. Accessed on: 16 Sep. 2022.
[19] Treves, T., The convention for the suppression of unlawful acts against the safety of maritime navigation. *Sing. J. Int'l & Comp. L.*, **2**, p. 541, 1998.
[20] Gladkov, G., Zhuravlev, M. & Katolikov, V., Hydromorphological substantiation of channel stability of navigable rivers in engineering water transport regulation of river runoff. *Energy Management of Municipal Transportation Facilities and Transport*, Springer: Cham, pp. 661–675, 2019. DOI: 10.1007/978-3-030-57453-6_63.
[21] Cheng, J.J., Xin, G.W., Zhi, L.Y. & Jiang, F.Q., Unloading characteristics of sand-drift in wind-shallow areas along railway and the effect of sand removal by force of wind. *Scientific Reports*, **7**(1), pp. 1–11, 2017. DOI: 10.1038/srep41462.
[22] *Kenya Railways v. Antares Co Pte Ltd* (The Antares) (No. 1), Lloyd's Rep 1 424, 1987.
[23] *Pyrene v. Scindia* case, 2 QB 402 (QBD), 1954.
[24] *Trafigura Beheer BV v. Golden Stavraetos Maritime Inc (The Sonia)*, E.W.C.A. Civ 664, 2003 Lloyd's Rep. 2 201, 2003.
[25] *Aries Tanker Corp v. Total Transport Ltd (The Aries)*, W.L.R.1 185, 1977.
[26] *Maxine Footwear Co Ltd v. Canadian Government Merchant Marine Ltd*, A.C. 589, 1959.
[27] *The Amstelstot* case, 2 Lloyd's Rep. 223 (HL), 1963.
[28] *The Muncaster Castle*, 1 Lloyd's Rep. 57 (HL), 1961.
[29] *The Yamatogawa* case, 2 Lloyd's Rep. 39 (QBD), 1990.
[30] *The Hellenic Dolphin* case, 2 Lloyd's Rep. 336 (QBD), 1978.
[31] *The El Greco (Australia) Ltd v. Mediterranean Shipping Co. SA*, FCA 588 (Au)., 2003.
[32] Berlingieri F., A comparative analysis of the Hague-Visby Rules, the Hamburg Rules and the Rotterdam Rules. https://comitemaritime.org/wp-content/uploads/2018/05/Comparative-analysis-of-the-Hague-Visby-Rules-the-Hamburg-Rules-and-the-Rotterdam-Rules-1.pdf. Accessed on: 16 Sep. 2022.
[33] *Reardon Smith Ltd. v. Black Sea and Baltic General Ins. Co*, 64 Ll. L. Rep. 229 (HL), 1939.
[34] *Stag Line v. Foscolo Mango*, A.C. 328 (HL), 1932.
[35] *The Al Taha* case, 2 Lloyd's Rep. 117 (QBD), 1990.
[36] Baatz, Y., Velasco, A.C., Debattista, C., Gurses, O., Hjalmarsson, J., Lista, A., Lorenzon, F., Pilley, R., Serdy, A., Shaw, R. & Tsimplis, M., *Maritime Law*, 4th ed., Taylor and Francis: London, 2020.
[37] *The Kapitan Sakharov*, 2 Lloyd's Rep. 255 (CA), 2000.

[38] Teare, N., Neutral Citation Number: [2020] EWHC 2604 (Admlty). https://www.quadrantchambers.com/sites/default/files/media/document/sakizaya_kalon_osios_david_v_panamax_alexander_2020_ewhc_2604_admlty.pdf. Accessed on: 16 Sep. 2022.
[39] Jaeung, C., Jinwoo, L., Changhee, L & Yulseong, K., Legal disputes under time charter in connection with the stranding of the *MV Ever Given*. *Sustainability*, **13**(19), p. 10559, 2021. DOI: 10.3390/su131910559.
[40] Christoph, B., *Ever Given*: An example of how complex international liability for damages can be. https://www.ibanet.org/ever-given-international-liability-damages. Accessed on: 16 Sep. 2022.
[41] Walton, L., The difference between Institute Cargo Clauses A, B and C. https://www.hayesparsons.co.uk/blog/the-difference-between-institute-cargo-clauses-a-b-and-c. Accessed on: 16 Sep. 2022.
[42] El Sawy, N., Lessons learnt from '*Ever Given*' incident in Suez Canal. https://www.thenationalnews.com/mena/2021/12/21/lessons-learnt-from-ever-given-incident-in-suez-canal/. Accessed on: 16 Sep. 2022.
[43] Gao, T. & Lu, J., The impacts of strait and canal blockages on the transportation costs of the Chinese fleet in the shipping network. *Maritime Policy and Management*, **46**(6), pp. 669–686, 2019. DOI: 10.1080/03088839.2019.1594423.

Author index

Agostinelli S. 113
Alkhereibi A. H. 37
Astiaso Garcia D. 113

Bakr A. F. 13
Bakri R. 13
Benitez I. 173
Boviatsis M. 185

de Juan M. 173
Duri B. 73

El Sakty K. 149
El-Sayad Z. 13

Gasparín P. 161
Ghosh G. 61

Islam A. E. 149

Li X. 87
Luke R. 73

Maghelal P. K. 87
Majidi Nezhad M. 113
Mandić N. 139
Marcos J. 173
Marrero Á. 161
Marseglia G. 25
Mesa J. A. 25
Miralles-Garcia J. L. 49
Musolino G. 123

Neshat M. 113

Ortega F. A. 25

Peda G. 123
Pehlevan Z. 99
Peronja I. 139
Piedra-de-la-Cuadra R. 25
Piras G. 113
Pourramazani H. 49

Qahtani A. 87

Ricci S. 99
Rodero F. 161
Russo F. 123

Sanz P. 161

Trumpold J. 3

Vukić L. 139

Wille C. 3

WIT Transactions on The Built Environment, Vol 212, © 2022 WIT Press
www.witpress.com, ISSN 1743-3509 (on-line)

Computers in Railways XVIII

Railway Engineering Design and Operation

Edited by: ***G. PASSERINI****, Marche Polytechnic University, Italy and* ***J.M. MERA****, Polytechnic University of Madrid, Spain*

Originating from papers presented at the 18th International Conference on Railway Engineering Design and Operation, this book provides up-to-date research on the use of advanced systems, promoting their general awareness throughout the management, design, manufacture and operation of railways and other emerging passenger, freight and transit systems. A key emphasis is placed on the use of computer systems in advanced railway engineering.

The included works are compiled from a variety of specialists interested in the development of railways, including managers, consultants, railway engineers, designers of advanced train control systems and computer specialists.

Topics covered include: Traffic safety, security and monitoring; Train and railways analysis; Operation of rail networks; Advanced train control; Energy-efficient design; Traffic modelling and simulation.

WIT Transactions on the Built Environment, Vol. 213
ISBN: 978-1-78466-479-4 eISBN: 978-1-78466-480-0
Published 2022 / 276pp

www.ingramcontent.com/pod-product-compliance
Lightning Source LLC
LaVergne TN
LVHW062238200926
848484LV00006B/1569